The Science of Waste

The Science of Waste

Frank R. Spellman

CRC Press
Taylor & Francis Group
Boca Raton London New York

CRC Press is an imprint of the
Taylor & Francis Group, an **informa** business

First edition published 2022
by CRC Press
6000 Broken Sound Parkway NW, Suite 300, Boca Raton, FL 33487-2742

and by CRC Press
2 Park Square, Milton Park, Abingdon, Oxon, OX14 4RN

CRC Press is an imprint of Taylor & Francis Group, LLC

Library of Congress Cataloging-in-Publication Data

Names: Spellman, Frank R., author.
Title: The science of waste / Frank R. Spellman.
Description: First edition. | Boca Raton, FL : CRC Press, 2022. | Includes bibliographical references and index.
Identifiers: LCCN 2021035301 (print) | LCCN 2021035302 (ebook) | ISBN 9781032172149 (hbk) | ISBN 9781032180601 (pbk) | ISBN 9781003252665 (ebk)
Subjects: LCSH: Refuse and refuse disposal. | Waste minimization.
Classification: LCC TD791 .S74 2022 (print) | LCC TD791 (ebook) | DDC 363.72/8--dc23
LC record available at https://lccn.loc.gov/2021035301
LC ebook record available at https://lccn.loc.gov/2021035302

ISBN: 978-1-032-17214-9 (hbk)
ISBN: 978-1-032-18060-1 (pbk)
ISBN: 978-1-003-25266-5 (ebk)

DOI: 10.1201/9781003252665

Typeset in Times
by Deanta Global Publishing Services, Chennai, India

Contents

Preface

As the title states, this is a book about waste—the science of waste (meaning "knowledge of" waste). Although important, this book is not about wasting electricity, wasting water, or burning sunshine, so to speak. In this book, waste is defined as something no longer wanted, something destroyed or broken or damaged beyond repair and therefore disposed of or simply thrown away because it is no longer functional, needed, and/or wanted. After providing a brief account of one example of waste in the United States during the enormous migration of people from the eastern portion of the country to the wide-open areas of the west, and after describing the different types of waste and waste management, the focus of this work turns to the question: Is waste really a waste? Stated differently, waste is not a waste if it can be recycled in some form or the other.

This book is intended to be used by environmental scientists, engineers, professionals, legal professionals, students, and instructors and is also designed to be used as a general reading resource for anyone who might want to obtain "knowledge of" waste and, when properly managed, its potential value to all of us.

This book is written to be understood without all gobbledygook and other spurious and non-essential information, commonly also known as gibberish. For those not familiar with any of the author's previous environmental works, it is important to point out that this book is written in the author's traditional conversation style. Failure to communicate is not an option; instead, it is a must, an obligation, and a commitment.

Frank R. Spellman
Norfolk, Virginia

Author

Frank R. Spellman is a retired assistant professor of Environmental Health at Old Dominion University, Norfolk, Virginia, and the author of over 155 books. Spellman has been cited in more than 400 publications; serves as a professional expert witness; incident/accident investigator for the U.S. Department of Justice and a private law firm; and consults on Homeland Security vulnerability assessments (VAs) for critical infrastructure including water/wastewater facilities nationwide. Dr. Spellman lectures on sewage treatment, water treatment, and homeland security and health and safety topics throughout the country and teaches water/wastewater operator short courses at Virginia Tech (Blacksburg, Virginia). He holds a BA in Public Administration; a BS in Business Management; an MBA; a Master of Science, MS, in Environmental Engineering; and a PhD in Environmental Engineering.

1 Waste by Any Other Name?

As described in the preface of this work, "waste" is defined as something no longer wanted, something destroyed or broken or damaged beyond repair, and therefore disposed of or simply thrown away because it is no longer functional, no longer needed, and/or no longer wanted. It is important to point out that "waste" is known by many names; many of them are described and discussed in this book. However, before listing the most common types of "waste" it is important to use an example of what waste can be when a product or material is not damaged but instead is necessary to dispose of. Consider, for example, the following factual account of when something treasured or wanted is discarded.

FRONTIER MENTALITY

It was one of the most famous trails ever. It was the longest overland trail in North America. In 1843, Americans were encouraged by the U.S. Government to travel the 2,000 miles from Independence, Missouri, to their final destination in the Oregon Territory to homestead the land. By settling the land with more Americans than British, Oregon would belong to the United States.

The 2,000-mile trek to Oregon Territory was no walk in the woods; it was a Great Migration of hordes of people traveling in covered wagons over mostly uncharted lands and trails (or so this was the case early on). The seemingly endless lines of prairie schooners snaked their way along the trail for more than six months before they arrived at their destinations.

The pioneers took what they could carry in their wagons. For example, it was not unusual for each wagon to be loaded with food that included: yeast for baking, crackers, cornmeal, bacon, eggs, dried meat, potatoes, rice, beans, and a big barrel of water. They would also take a cow if they had one. Pioneers made their own clothing, so they brought cloth to sew, needles, threads, pins, scissors, and leather to fix worn-out shoes. They had to make their own repairs, so they brought saws, hammers, axes, nails, strings, and knives.

Occasionally, and against the advice of the wagon master, pioneers would also pack away personal treasures—heirlooms such as pianos, family trunks and furniture, mirrors, assorted chinaware, silver, paintings, and other decorative household goods of the day. These items not only took up a lot of space within the wagon but also were clumsy to handle and added extra weight to the wagon.

It usually did not take too many miles before many of those who had packed too much started to drop off or discard various personal treasures along the trail. So many personal property items were discarded along the trail that by the late 1840s it

DOI: 10.1201/9781003252665-1

1

was no longer necessary to follow the tracks of the wagons that had preceded them to follow and stay on the trail. All one needed to do was to follow the strewn and rotting personal treasures (and assorted gravesites) that marked the trail. The parting with their personal treasures must have been heartbreaking to many of the pioneers as they made the difficult trek along the Oregon Trail to their new homesteads in the West.

For decades many have asked what was it that drove or enticed these people to undertake such a perilous adventure into lands unknown. The main draw, of course, was the promise of free land. For others it was freedom from the squalor of eastern city life or the drudgery of farm life in trying to eke out a living on worn-out soils as tenet farmers that drove them to undertake the arduous Western adventure. For others it was the wide-open untamed spaces and the quest for adventure that was the drawing card.

This burning desire to conquer new lands, exploit them, and become rich was like a powerful magnet attracting and aligning the metal filings of their mentalities. This outlook, mode, or way of thought that drove the pioneers to trek across barren, rugged, unforgiving wilderness is commonly called frontier mentality. That is, we live as though we can't effectively harm the natural world in a significant way because it is so big and we are so little, and if we damage one place, there will always be a new frontier to move to. The reality is that this mode of thought, this frontier mentality, can be summed up today by simply stating that in the United States, there has always been a Western frontier, a place to go to and start over where there are riches just for the taking for hardy spirit and determined worker.

Let's get back to the pioneers' trek along the Oregon Trail where they have marked the trail with ruts carved by their wagon wheels and various other signposts provided by their discarded personal goods. Again, after years of traversing the Oregon Trail, wagon train after wagon train discarded goods that dotted the prairie areas for miles. Humans have this tendency; that is, when some object they own is no longer needed or no longer pleases them or has outlived its usefulness they simply discard it—out of sight, out of mind. The pioneers' frontier mentality about the west being an expanse of wide-open untamed space gave them no qualms about leaving their personal goods to rot along the trail—no qualms about polluting the landscape.

This same frontier mentality—"I no longer want or need it, so I'll throw it away"—did not stop when the pioneers reached their destinations. The mindset alters a bit, but stops? No. It began with settlements which were almost exclusively built along rivers. The rivers provided a convenient means of disposing the unwanted. Beginning with simple discards such as coffee grounds to more complex items such as white goods (washers, dryers, refrigerators, etc.), to even more complex and persistent chemical compounds and mixtures, the river was the repository of choice for all. The thinking was, of course, that no matter what you threw into the river the running water would purify itself every 10 miles or so. But, when there are several settlements with hundreds or thousands of people up and down the length of the river, the purification capacity of the running water is exhausted. However, even though we are running out of pristine areas to pollute, our throw-away society continues to be strongly influenced by our consumerism and excessive disposal of short-lived items.

WASTE BY OTHER NAMES

One thing is certain; there are not only several different forms of waste and they are all categorized in a variety of ways, but they also are identified by their own designation or name. In this section, a representative sample of the major types of waste listed and described here is not necessarily exclusive, and keep in mind that there can be (and is) considerable connection and/or overlap so that one waste entity may fall into one to many types, forms, and/or categories.

TYPE AND WASTE CHARACTERIZATION

- **Agricultural waste**—consists of unwanted, useless, undesirable, or unsalable substances and materials produced wholly from agricultural procedures and operations directly related to the growing of crops or raising of animals for the most important purpose of making a profit or for a source of revenue. A few examples of agricultural waste include fruit-bearing trees, vegetables, grape vines, and date palm fronds. On the other hand, common wastes such as plastic, trash, garbage, tires, rubber, oil filters, pallets, chemically treated wood, asbestos-containing materials (ACM), fertilizer and pesticide waste, construction and demolition materials, packaging material and boxes, broken boxes, and grass, weeds, and tree trimmings are not classified as agricultural waste. Also note that in an agricultural location such as a vineyard and/or an orchard, where waste is produced because of land use conversion to nonagricultural purposes, the waste produced is not classified as agricultural waste.
- **Animal byproducts (ABP)**—consists of animal carcasses, parts of animal carcasses, or animal parts not suitable for human consumption. ABP also includes dairy shed effluent containing urine, dung, wash water, residual milk, dairy manure, spilled feed, and other wastes including bedding materials, feathers, wood chips, straw hay, and other sources of organic waste.
- **Biodegradable waste**—consists of any waste containing organic matter which can be broken down into carbon dioxide, water, methane, or simple organic molecules by microorganisms and other living things by composting, aerobic digestion, anaerobic digestion, or similar processes. There are also a few inorganic materials that are biodegradable such as wall board and associated simple organic sulfates; these are decomposed in anaerobic landfills to hydrogen sulfide.
- **Biomedical waste**—consists of any kind of waste containing infectious materials. It may also include waste such as packaging, unused bandages, and infusion kits that is associated with the generation of biomedical waste that visually appears to be of medical or laboratory or research laboratory origin. Discarded sharps, used or not used, are considered biomedical waste.
- **Bulky waste**—consists of waste types that are too large to be accepted by the regular waste collection and generally includes furniture, large appliances, and plumbing fixtures. In many locations, branches, brush, roots, and other green waste are categorized as bulky waste.

- **Business waste**—consists of commercial waste and industrial waste.
- **Chemical waste**—consists of waste from harmful chemicals.
- **Clinical waste**—see biomedical waste.
- **Coffee waste (aka coffee effluent)**—consists of a byproduct of coffee processing.
- **Commercial waste**—consists of waste from premises used mainly for the purposes of business or trade and also for education, recreation, entertainment, and sports.
- **Composite waste**—consists of waste made up of two or more materials.
- **Concentrated animal feeding operations (CAFOs)**—CAFOs are farming operations where large numbers (often in the thousands of animals) of livestock or poultry are housed inside buildings or in confined feedlots. How many animals? The USEPA defines a CAFO or industrial operation as a concentrated animal feeding operation where animals are confined for more than 45 days per year. To classify as a CAFO, such an operation must also have over 1,000 animal units—a standardized number based on the amount of waste each species produces, basically 1,000 pounds of animal weight. Thus dairy cattle count as 1.4 animal units each. A CAFO could house more than 750 mature dairy cattle (milking and or dry cows) or 500 horses and discharge into navigable water through a man-made ditch or a similarly man-made device. CAFO classification sets numbers for various species per 1,000 animal units:
    - 2,500 hogs
    - 700 dairy cattle
    - 1,000 beef cattle
    - 100,000 broiler chickens
    - 82,000 layer hens
- **Construction and demolition waste**—construction waste consists of debris from the construction process. Demolition waste consists of debris from the destruction of buildings, roads, bridges, or other structures.
- **Consumable waste**—consists of wastes from common goods typically used within a year.
- **Controlled waste**—consists of waste that is subject to legislative control in either its handling or its disposal.
- **Dog waste**—consists of solid wastes from dogs.
- **E-waste**—consists of discarded electrical or electronic devices.
- **Food waste**—consists of food that is not eaten and thrown away.
- **Green waste**—consists of any organic waste that can be composted.
- **Grey water**—consists of all the wastewater generated from office buildings or streams without fecal contamination.
- **Hazardous waste**—consists of waste that has substantial or potential threats to public health or the environment.
- **Human waste**—consists of the waste products of the human digestive system and the human metabolism, specifically urine and feces.
- **Industrial waste**—consists of waste produced by industrial activity which includes any material that is rendered useless during a manufacturing process such as that of mill, mining, and factory operations.

- **Inert waste**—consists of waste which is neither chemically nor biologically reactive and will not decompose or only very slowly (e.g., sand and concrete).
- **Inorganic waste**—consists of waste that does not contain organic compounds (e.g., glass, aluminum containers, and dust).
- **Litter**—consists of waste products that have been thrown away improperly, without consent, in an unsuitable location.
- **Marine debris (aka marine litter)**—consists of human-created waste that has deliberately or accidentally been released into a sea or ocean.
- **Metabolic waste**—consists of substances left over from metabolic processes—such as cellular respiration—which cannot be used by the organism because they are surplus or toxic and must therefore be excreted.
- **Mixed waste**—consists of commercial and municipal wastes that are mixtures of plastics, metals, glass, and biodegradable waste such as paper and textiles.
- **Municipal solid waste (MSW)**—consists of what is more commonly known as trash or garbage—consists of everyday items we use and then throw away, such as product packaging, grass clippings, furniture, clothing, bottles, food scraps, newspapers, appliances, paint, and batteries.

DID YOU KNOW?

With regard to MSW, in 2013, newspapers/mechanical papers recovery was about 67% (5.4 million tons) and about 60% of yard trimmings were recovered. Organic materials continue to be the largest component of MSW. Paper and paperboard account for 27% and yard trimmings, and food accounts for another 28%. Plastics comprise about 13%; metals make up 9%; and rubber, leather, and textiles account for 9%. Wood follows at around 6%. And glass at 5%. Other miscellaneous wastes make up approximately 3% of the MSW generation in 2013 (USEPA, 2016).

- **Packaging waste**—consists of containers and packaging products.
- **Post-consumer waste**—consists of a waste produced by the end consumer of a material stream.
- **Radioactive waste**—consists of a type of hazardous waste that contains radioactive material.
- **Sewage**—consists of a type of wastewater that is produced by a community of people.
- **Sharps waste**—consists of a form of biomedical waste composed of used devices used to puncture or lacerate the skin.
- **Slaughterhouse waste**—consists of waste derived from the body of an animal.
- **Transuranic waste**—consists of wastes that are artificially made; radioactive elements, such as neptunium, plutonium, americium, and others.

FOREVER RELICS

What are forever relics?

Forever relics are leftovers, remnants, remainders, residues, artifacts, remains, or historical objects.

Okay, so what do forever relics have to do with waste?

Everything.

One of the characteristics of waste is that it has a long lifespan. Archaeologists, anthropologists, historians, and others, for example, take advantage of this characteristic. Wastes found in old dumps, called *middens*, provide a useful resource for study; they are useful relics that illustrate what the diets and habits of past societies were about. Wastes consisting of animal bones, human excrement, botanical material, mollusk shells, sherds (e.g., pottery fragments), lithics (i.e., artifacts made of stone), and other artifacts and ecofacts (e.g., animals bones and plant pollen) were associated with past human occupation.

DID YOU KNOW?

The word "midden" is still in use in Scotland and refers to anything that is a mess, a muddle, or chaos (SLC, 2020).

Midden piles provide the perfect environment to preserve debris (even organic remains) of daily life in the past. The anaerobic conditions ensure the main natural decomposer, oxygen, is prohibited from entering the midden. Each component of the midden is an individual component combining to form a different mix depending on the activity associated with the particular deposit. During deposition, sedimentary material is also deposited. Different mechanisms, from wind and water to animal digs, create a matrix which can also be analyzed to provide climatic and seasonal information.

THE BOTTOM LINE

Waste provided by humans in whatever form and/or wherever stored will continue to reveal to those who study the past a layered record of who we were and how we lived at a specific time in the past. Relics of the past have told us much and those of the present and future may continue a catalogue of how we live and lived.

REFERENCES

SLC (Scots Language Centre). (2020). Annaker's midden n. a mess, a shambles. Accessed 1/25/2021 @ https://scotslanguage.com/articles/view/id/4494.
USEPA. (2016). Municipal solid waste. Accessed 1/27/2021 @ https://archive.epa.gov/epawaste/nonhaz/municipal/web/html/index.htm.

2 Cave to Allegorical Cave to Present

Salt, meet wound. Insult, greet injury. Waste, say hello to the environment.

—**Frank R. Spellman**

THE CAVE

By today's calendar, it is 26 June 15,543 BC. The place, a large natural cave set deep under a solid outcropping that formed a fairly significant mountain meadow before the last glacial ice-sheet gouged, gorged, ground, and pulverized it down to its present size and shape.

The colossal sheet of ice is in retreat. When it was at its full width, depth, and length, it extended several hundred miles beyond the cave site to a V-shaped valley, where glacial melt fed a youthful, raging river that ran through the valley's bottomland.

A small, steady stream of melt-water courses almost in a straight line past one side of the cave, down toward that valley, where it joins and feeds that river.

On the other side is a sloping field of young grass, brush, and flowers—flowers everywhere. Up close, we see the stark remnants of the terminal moraine that formed this abrupt slope with its fresh cover of grass and blossoms. A closer look reveals a dark heap at the base of the slope—a heap of trash: skin, sinew, bone, decaying corpses, and burnt remnants of past hunts and feasts. We know only too well the refuse, filth, and discards that people leave behind as their foulest signature; somebody, maybe many bodies live close by. Where? Of course—in the cave. Let's take a look inside.

We wander up to the huge hole in the rock that forms the mouth of the cave. We tread carefully—we do not want to disturb (startle, frighten—anger) the occupants. Remember, we're talking about caveman here. No language in common. No culture in common. Could we have anything in common with such … primitive people? But all is quiet—and with no overt threat present, our curiosity overcomes our caution, and walk into the opening chamber. And something reaches out and grabs us—not caveman, but a stench—a horrible stench—too horrible to describe. With our fingers clamped tightly to our noses, crushing to the bone, we move on, too interested, despite the reek, to retreat.

We can see fairly well in this chamber because daylight pours in through the entrance. We take a few steps and stop to look around, religiously not breathing through our noses. The walls of the cave are covered with black soot. A pit near the cave wall to the right, under attack by millions of flies and other insects, provides

DOI: 10.1201/9781003252665-2

much of the stench—the latrine. A heap of detritus similar to the dump outside provides the rest of the reek. That this cave is abandoned dawns on us. We have no doubt as to why. The largest by-product manufactured by mankind has taken over—the cave is a garbage dump.

Perhaps, back in other chambers, deeper in the cave, exist cave paintings and remnants of ceremony. But we don't have the tools to explore them with us today, and we retreat, grateful for a breath of fresh air.

Outside, a few hundred feet from the smelly cave, and within sight of the garbage heap, we stop to contemplate what we've seen. Fifteen thousand years from now, archaeologists will find this cave, and explore it thoroughly, learning information that will give us insights into the world of the people whose former home we have just visited. But the remains the archaeologists will find will be altered by 15,000 years of history. The picture they see will be incomplete, scattered by the natural interferences life causes, giving mystery to the short and brutal lives of our ancestors.

But here, right now, we see similarities. We foul our environment in the same ways, and in more. But the caveman had a huge advantage over modern man, in that respect. When his living quarters became too foul for comfort, he could pick up whatever he considered of value he and his tribe could carry and move on. A fresh site was always just around the next bend in the river. The pollution he created was completely (eventually) naturally biodegradable—in a few years, this cave could house humans again.

Although we have our similarities with those far-off ancestors, one stark difference is plain: modern man cannot destroy and pollute his environment with impunity. We can no longer simply pick up stakes and move on. What we do to our environment has ramifications on a scale that we cannot ignore—or avoid.

THE ALLEGORICAL CAVE

The allegorical cave? What is it and what does it have to do with waste? Well, the preceding presentation detailing waste accumulation in ancient cave-persons' home makes the point that the generation of waste is a historical problem. From the first moment humans or human-like beings walked upon Earth, waste has been a companion all the way through the changing milieus—although a short-lived companion, for sure. The attitude from the beginning has been I do not want or need it any longer so throw it out; get rid of it; the old out of sight out of mind scenario.

Again, what does allegory of the cave have to do with waste?

Let's start at the beginning and the connection will be as clear as a mountain stream, a clear day and/or as clear as one's empty waste can.

I am referring to Plato's *Allegory of the Cave*. If you are not familiar with this circa 380 BCE excerpt from Plato's classic work *The Republic*, the following will provide you with a brief overview of Plato's work from his *Republic*, VII 514 a, 2 to 517 a, 7.

Note: This is the author's rendition of Plato's classic. In this presentation, Socrates is speaking to Glaucon the older brother of Plato. The Allegorical Cave is an adaption from Jowett (2020).

THE SETTING: THE CAVE

People are living under the earth in a cave dwelling. Stretching a long way up toward the daylight is its entrance, toward which the entire cave is amassed. The people have resided in the cave since childhood, constrained by the legs and neck. Thus, they stay in the same place so that there is only one thing for them to look at; that is, they only see what they encounter in front of their faces. They are unable to turn their heads around due to the constraints; you could say that they are fettered in place, for sure.

THE FIRE

The constrained people are allowed some light, namely from a fire that casts its glow toward them from behind them, being above and at some distance. Between the fire and those who are constrained (i.e., behind their backs) there runs a walkway at a certain height. A low wall has been built the length of the walkway, like the low curtain that puppeteers put up, over which they show their puppets.

IMAGES FROM THE FIRE

All along the long wall, people are carrying all sorts of things that reach up higher than the wall: statues and other carvings made of wood or stone and many other artifacts that people have made. Some of the people are talking to each other as they wall along and some are quiet.

In a discussion between Socrates and the student Glaucon the student says, "this is an unusual picture presented here, and these are atypical prisoners."

"They are very much like us," Socrates responds.

WHAT THE PRISONERS SEE AND HEAR

"From the beginning people like these have never managed, whether on their own or with the help of others, to see anything besides the shadows that are always projected on the wall opposite them by the glow of the fire," says Socrates.

Glaucon says, "How could it be otherwise, as they are constrained to keep their heads fixed for their entire lives?"

"And what do they see of the things that are being carried along behind them? Do they not simply see the shadows?" replies Socrates.

"Without doubt," replies Glaucon.

"Well now if they were able to say something about what they saw and to talk it over, do you not think that they would consider that which they saw on the wall as people?" Socrates asks.

Glaucon says, "they would have to."

"And now what if this prison also has an echo reverberating off the wall in front of them? Whenever one to the people carrying the things and walking behind those in restraints would make a sound, do you think the prisoners would imagine that the speakers were anyone other than the shadow passing in front of them?" Socrates asks.

"Nothing else," replies Glaucon.

"Well, all in all, those who were restrained would consider nothing besides the shadows of the artifacts as the unhidden," Socrates says.

"That would have to be true," says Glaucon.

A Prisoner Gets Free

Socrates says, "Watch the process whereby the prisoners are set free from their restraints and also cured of their lack of insight, and also consider what kind of lack of comprehension must be if the following were to happen to those who were restrained."

Walks Back to the Fire

Socrates says, "Whenever any of them was unrestrained and was coerced to stand up suddenly, to turn around, to walk, and to look up toward the light, in each case the person would be able to do this only with pain and because of the flickering brightness would be unable to look at those objects whose shadows he up to that time saw."

Is Questioned about the Objects

Socrates says, "If all this were to happen to the prisoner, what do you think he would say if someone were to advise him that what he saw were ordinary trivialities but that now he was nearer to persons; and that, as a result of now being turned toward what is more in being, he also saw more accurately?"

The Answer He Gives

Socrates says, "And then if someone were to show him any of the things that were passing by and coerced him to answer the question about what it was, don't you think that he would be at in a state of distress and in addition would consider that what he previously saw with his own eyes was more visible than what was now being revealed to him by someone else."

Glaucon answers, "Yes, for sure."

Looking at the Fire-light Itself

Socrates says, "And if someone even coerced him to look into the glare of the fire, should his eyes not hurt him, and would he not then turn away and bolt back to that which he is capable of looking at? And would be not decide that was in fact clearer than what was now being shown to him?"

Glaucon says, "Exactly."

Out of the Cave into Daylight with Pain, Rage, Blindness

Socrates says, "However, if someone, using force, were to pull him away from there and to drag him up the cave's rough and steep ascent and not to let go of him until he had dragged him out into the light of the sun would not the one who have been dragged like this feel, in the process, pain and rage? And when he got into the sunlight, would his eyes be filled with the glare, and wouldn't he thus be unable to see any of the things that are now revealed to him as the unhidden?"

Glaucon replies, "He would not be able to that, at least not right way."

GETTING USED TO THE LIGHT

Socrates says, "It would obviously take some time for his eyes to become acclimated to the sun."

SHADOWS AND REFLECTIONS

Socrates says, "In the process of acclimatization he would first and most easily be able to look at shadows and after that the images of people and the rest of things as they are reflected in water."

LOOKING AT THINGS DIRECTLY

Socrates says, "However, later he would be able to view the things themselves—the actual beings and not the reflections. But within the span of such things, he might well envision what there is in the heavenly arena, and this arena itself, more easily during the night by looking at the light of the stars and the moon than by looking at the sun and its glare during the day."

Glaucon replies, "Absolutely."

LOOKING AT THE SUN ITSELF

Socrates says, "I think that finally he would be in a condition to look at the sun itself, not just at its reflection whether in water or wherever else if might appear."

Glaucon replies, "Yes, it would happen this way."

THOUGHTS ABOUT THE SUN: ITS NATURE AND FUNCTIONS

Socrates says, "And having done all that, by this time he would also be able to gather the following about the sun: First, that it is that which gives both the seasons and the years; second, it is that which controls whatever there is in the now visible region of sunlight; and third that it is also the origin of all those things that the people dwelling in the cave have before their eyes in some way or other."

Glaucon replies, "It seems obvious that he would get to these things after he had gone out beyond those previous things, those mere shadows and reflections."

THOUGHTS ABOUT THE CAVE

Socrates says, "And then what? If he again recalled his first dwelling, and the "knowing" that passes as the norm there, and the people with whom he once was restrained, don't you think he would consider himself fortunate because of the transformation that had occurred and, by comparison, feel sorry for them?"

Glaucon replies, "Yes, he would."

WHAT COUNTS FOR "UNDERSTANDING" IN THE CAVE

Socrates says, "On the other hand, what if among the people in the previous dwelling, the cave, certain honors and commendations were established for whomever most clearly catches sight of what passes by and also best remembers which of them normally is brought by first, which one later, and which ones at the same time? And what if there were honors for whoever could most easily anticipate which one might come by next?"

WHAT WOULD THE LIBERATED PRISONER NOW PREFER?

Socrates says, "Do you think the one who had gotten out of the cave would still covet those within the cave and would want to contend with them who are esteemed and who have power? Or would not he or she much rather wish for the condition that Homer speaks of, namely 'to live on the land above ground as the paid menial of another destitute peasant'? Wouldn't he or she prefer to put up with absolutely anything else rather than associate with those opinions that hold in the cave and be that kind of human being?"

Glaucon replies, "I think that he or she would prefer to endure everything rather than be that kind of human being."

THE RETURN: BLINDNESS

Socrates says, "And now, I responded, consider this: If this person who had gotten out of the cave were to go back down again and sit in the same place as before, would he not find in that case, coming suddenly out of the sunlight, that his eyes are filled with darkness?"

Glaucon replies, "Yes, very much so."

THE DEBATE WITH THE OTHER PRISONERS

Socrates says, "Now if once again, along with those who had remained constrained there, the freed person had to engage in the business of asserting and maintaining opinions about the shadows—while his eyes are still weak and before they have readjusted, an adjustment that would require quite a bit of time—would he not then be exposed to ridicule down there? And would they not let him know that he had gone up but only in order to come back down into the cave with his eyes ruined—and thus it certainly does not pay to go up."

AND THE FINAL OUTCOME

Socrates says, "And if they can get hold of a person who takes it in hand to free them from their chains and to lead them up, and if they could kill him, will they not actually kill him?"

Glaucon replies, "Absolutely."

After perusing Plato's *The Allegory of the Cave*, the obvious question is what does it have to do with waste? Does it mean that society is forced into a cave of misunderstanding? Does it mean that the occupants restrained in the modern cave are blind to the truth? Does it have any bearing at all on our waste problems of today?

All good questions. For the purposes of this text, Plato's allegory points out that many of us hold or possess idiosyncratic biases—"it's just waste, I don't need it anymore … please throw it away … put it anywhere but here … let someone else handle it, live with it, dispose of it." Another problem we have today with regard to waste and its production and disposal is the preoccupation of individuals. For example, when thinking about something, anything else while driving a vehicle on the Interstate and when we open the car, truck, RV window and throw out the expended cigarette, the gum wrapper, the candy wrapper, the rest of the sandwich, the soda bottle, or anything else we do not want anymore this occurs because we are preoccupied with our thoughts and do not even give a micro-moment's thought to what we are doing with the waste. Waste, it is only waste. The actual forms are out the window we can't know or think about them directly and venture toward the light using reason.

The bottom line: We do not comprehend, because we are preoccupied that waste does not go away.

FAST FORWARD FROM THE PAST TO THE PRESENT

SALMON AND THE RACHEL RIVER*

The Rachel River, a hypothetical river system in the northwestern United States, courses its way through an area that includes a Native American Reservation. The river system outfalls to the Pacific Ocean and the headwaters begin deep, remote, and high within the Cascade Mountain Range of Washington State. For untold centuries, this river system provided a natural spawning area for salmon. The salmon fry thrived in the river, and eventually grew the characteristic dark blotches on their bodies and transformed from fry to parr. When the time came to make their way to the sea, their bodies now larger and covered with silver pigment, the salmon, now called smolt, inexorably migrated to the ocean, where they thrived until time to return to the river and spawn (about 4 years later). In spawning season, the salmon instinctively homed their way toward the odor generated by the Rachel River (their homing signal), and up the river to their home waters, as their life cycle instincts demand.

Before non-Native Americans (settlers) arrived in this pristine wilderness region, nature, humans, and salmon lived in harmony and provided for each other. Nature gave the salmon the perfect habitat; the salmon provided Native Americans with

---

* *Salmon and the Rachel River* is based on and adapted from F.R. Spellman (2006) *Environmental Science and Technology: Concepts and Applications*, 2nd edition. Rockville, Maryland: Government Institutes.

sustenance. Native Americans provided to both their natural world and the salmon the respect they deserve.

After the settlers came to the Rachel River Valley, changes began. The salmon still ran the river, humans still fed on the salmon, but circumstances quickly altered. The settlers wanted more land, and Native Americans were forced to give way; they were destroyed or forcibly moved to other places, to reservations, where the settlers did all they could to erase Native American beliefs and cultural inheritance. The salmon still ran the streams.

After the settlers drove out the Native Americans, the salmon continued to run, for a while. But more non-Native Americans poured into the area. As the area became more crowded, the salmon still ran, but now their home, their habitat, the Rachel River, started to show the effects of modern civilization's influence. The "civilized" practice and philosophy was "If I don't want it any more, it's trash. Throw it away ... it's just waste," and the river provided a seemingly endless dump—out of the way, out of sight, out of mind. And they threw their trash away, all the mountains of trash they could manufacture, into the river. The salmon still ran.

More time passed. More people moved in, and the more people, the bigger their demands. In its natural course, sometimes the river flooded, creating problems for the settler populations. Besides, everyone wanted power to maintain modern lifestyles—and hydropower poured down the Rachel River to the ocean constantly. So they built flood control systems and a dam to convert hydropower to hydroelectric power. (Funny; the Native Americans didn't have a problem with flood control. When the river rose, they broke camp and moved to higher ground. Hydroelectric power? If you don't build your life around things, you don't need electricity to make them work. With the sun, the moon, and the stars and their healthy, vital land at hand, who would want hydroelectric power?)

The salmon still ran.

Building dams and flood control systems takes time, but humans, though impatient, have a way of conquering and using time (and anything else that gets in the way) to accomplish their tasks, goals, objectives—and construction projects. As the years passed, the construction moved on to completion, and finally ended. The salmon still ran—but in reduced numbers and size. Soon local inhabitants couldn't catch the quantity and quality of salmon they had in the past. When the inconvenience finally struck home, they began to ask, "Where are the salmon?"

But no one seemed to know. Obviously, the time had come to call in the scientists, the experts. So the inhabitant's governing officials formed a committee and funded a study and hired some scientists to tell them what was wrong. "The scientists will know the answer. They'll know what to do," they said, and that was partly true. Notice they didn't try to ask the Native Americans. They also would have known what to do. The salmon had already told them—water is life's mother. There is no life without water.

The scientists came and studied the situation, conducted testing, tested their tests, and decided that the salmon population needed to increase. They determined increased salmon population could be achieved by building a fish hatchery, which would take the eggs from spawning salmon, raise the eggs to fingerling-sized fish, release them into specially built basins, and later, release them to restock the river.

A lot of science goes into the operation of a fish hatchery. It can't operate successfully on its own (though Mother Nature never has a serious problem with it when left alone) but must be run by trained scientists and technicians following a proven protocol based on biological studies of salmon life cycles.

When the time was right, the salmon were released into the river—meanwhile, other scientists and engineers realized that some mechanism had to be installed in the dam to allow the salmon to swim downstream to the ocean, and the reverse, as well. In salmon lives (since they are an anadromous species—they spend their adult lives at sea but return to freshwater to spawn), what goes down must go up (upstream). Those salmon would eventually need some way of getting back up past the dam and into home water, their spawning grounds. So the scientists and engineers devised, designed, built, and installed fish ladders in the dam so that the salmon could climb the ladders, scale the dam, and return to their native waters to spawn and die.

In a few seasons, the salmon again ran strong in the Rachel River. The scientists had temporarily—and at a high financial expenditure—solved the problem. Nothing in life or in Nature is static or permanent. All things change. They shift from static to dynamic, in natural cycles that defy human intervention, relatively quickly, without notice—global climate change, a dormant volcano, or the Pacific Rim tectonic plates. In a few years, local Rachel River residents noticed an alarming trend. Studies over a five-year period showed that no matter how many salmon were released into the river, fewer and fewer returned to spawn each season.

So they called in the scientists again. "Don't worry. The scientists will know. They'll tell us what to do."

The scientists came in, analyzed the problem, and came up with five conclusions:

1. The Rachel River is extremely polluted both from point and nonpoint sources. The causal factor(s): wastes from improper garbage disposal including runoff of pesticides and fertilizers and from animal wastes from farmlands into the river.
2. The Rachel River Dam has radically reduced the number of returning salmon to the spawning grounds.
3. Foreign fishing fleets off the Pacific Coast are depleting the salmon.
4. Native Americans were removing salmon downstream, before they even get close to the fish ladder at Rachel River Dam.
5. A large percentage of water is withdrawn each year from rivers for cooling machinery in local factories. Large rivers with rapid flow rates usually can dissipate heat rapidly and suffer little ecological damage unless their flow rates are sharply reduced by seasonal fluctuations. This was not the case, of course, with the Rachel River. The large input of heated, wasted water from Rachel River area factories back into the slow-moving Rachel River creates an adverse effect called **thermal pollution**. Thermal pollution and salmon do not mix. In the first place, increased water temperatures lowers

the dissolved oxygen (DO) content by decreasing the solubility of oxygen in the river water, and warmer river water also causes aquatic organisms to increase their respiration rates and consume oxygen faster, increasing their susceptibility to disease, parasites, and toxic chemicals. Although salmon can survive in heated water—to a point—many other fish (the salmon's food supply) cannot. Heated discharge water from the factories also disrupts the spawning process and kills the young fry.

The scientists prepared their written findings and presented them to city officials, who read them and were (at first) pleased. "Ah!" they said. "Now we know why we have fewer salmon!"

But their short-lived pleasure faded. They did indeed have the causal factors defined—but what was the solution? The scientists looked at each other and shrugged. "That's not my job," they said. "Call in the environmental folks."

The salmon still ran, but not up the Rachel River to its headwaters.

Within days, the city officials hired an environmental engineering firm to study the salmon depletion problem. The environmentalists came up with the same causal conclusions as the scientists (which they also related to the city official), but they also related the political, economic, and philosophical implications of the situation to the city powers. The environmentalists explained that most of the pollution constantly pouring into the Rachel River would soon be eliminated when the city's new wastewater treatment plant came on line, and that particular **point source pollution—a serious waste source**—was eliminated. They explained that the state agricultural department and their environmental staff were working with farmers along the lower river course to modify their farming practices and pesticide treatment regimens to help control the most destructive types of **nonpoint source pollution**. The environmentalists explained that the Rachel River dam's present fish ladder was incorrectly configured but could be modified with minor retrofitting.

They explained that the over-fishing by the foreign fishing fleets off the Pacific Coast was a problem that the federal government was working to resolve with the governments involved. The environmentalists explained that the State of Washington and the federal government were addressing the problem with the Native Americans fishing the down-river locations before the salmon ever reached the dam. Both governmental entities were negotiating with the local tribes on this problem, and the local tribes had pending litigation against the state and the federal government on who actually owned fishing rights to the Rachel River and the salmon.

For the final problem, the thermal waste stream from the factories making the Rachel River unfavorable for spawning, decreasing salmon food supply, and/or killing off the young salmon fry, the environmentalists explained that to correct this problem, the outfalls from the factories would have to be changed—relocated. The environmentalists also recommended the construction of a channel basin whereby the ready-to-release salmon fry could be released in a favorable environment, at ambient stream temperatures, and would have a controlled one-way route to safe downstream locations where they could thrive until time to migrate to the sea.

After many debates and many newspaper editorials, city officials put the matter to a vote—and voted to fund the projects needed to solve the salmon problem in the Rachel River. Some short-term projects are already showing positive signs of change, long-term projects are underway, and the Rachel River is on its way to recovery.

In short, scientists are professionals who study to find "the" answer to a problem through scientific analysis and study. Their interest is in pure science. Environmental engineers and scientists can arrive at the same causal conclusions as the pure scientists but are also able to factor in socio-economic, political, and cultural influences as well.

But wait! It still isn't over. Concerns over the disruption of the wild salmon gene pool by hatchery trout are drawing attention from environmentalists, conservationists, and wildlife biologists. Hatchery- or farm-raised stock of any kind is susceptible to problems caused by, among other things, a lack of free genetic mixing, the spread of disease, infection, and parasites, reinforcement of negative characteristics—when escaped hatchery salmon breed with wild salmon, the genetic strain is changed, diseases can be transmitted … many problems arise—pointing out that fixing an environmental problem is not always the final word on the problem.

The bottom line: Waste and waste-producing activities do not go away. You might say that the problem with waste when it is mitigated simply morphs into other problems. Again, waste does not go away. Instead, means must be found and employed to transform the waste into something useful, less harmful. However, the better solution is not to produce any waste in the first place. Is it possible to produce substances and products that do not produce waste?

REFERENCE

Jowett, B. (2020). *The Republic*. NY: Vintage Classics.

3 Waste?
It's in the Garbage Can

Do you suffer from white nights when sleep is somewhere else, elusive, non-existent?—worrying yourself sick and silly about Drexler's (1986) grey goo; you know, that end-of the-world scenario involving molecular nanotechnology in which out-of-control self-replicating machines (robots) consume all matter on Earth while building more of themselves. Or that you might be buried and suffocated in the waste they produce? An important point to keep in mind with regard to the technology and environment connection is that, generally speaking, it is not the old or new technologies that may adversely affect the environment; instead, it is usually the processes used to manufacture the technology and the waste stream generated during the production process that adversely affect the environment.

—**F. R. Spellman**

SETTING THE RECORD STRAIGHT

"Have you heard anyone say that waste is not an issue a problem because it is in the garbage can, exactly where it belongs? It is gone … not my problem anymore."

So that is true, isn't it? Once the trash (dry items), garbage (wet items), and waste or rubbish as a general term include expressions for all discards in the garbage can or receptacle, it is someone else's problem, right?

Wrong. Wastes of any type do not go away; instead, they are just moved from one location to another or turned from one substance into another. This is not to say that this is all bad. When waste is turned into a reusable substance or product (recycled), it can be beneficial depending upon what the transformed substance or product's end use or status is.

The list of waste types presented earlier is extensive. It would be difficult to cover every waste type in one presentation. Therefore, this book presents a discussion on selected waste issues other than toxins, mining waste, nuclear waste, and farming waste and instead focuses on those wastes that are currently flagrant issues on a global scale. These wastes include municipal solid waste, electronic waste, litter, marine debris, medical waste, leachate, and incinerated wastes. Intermixed within this discussion of these current waste issues is a detailed examination of landfills, ocean garbage patches, and waste disposal incidents.

Before discussing these issues, it is important to point out that throwing garbage away, any type of garbage, is not a one-and-done proposition. Remember, the waste does not go away. Because of the growing problem with waste and its persistence,

DOI: 10.1201/9781003252665-3

the United States Environmental Protection Agency (USEPA) has come up with a Sustainable Materials Management (SSM) plan. Is this plan the cure-all for solving the waste problem? In other words, is SSM, the "silver bullet," needed to mitigate this growing problem? Well, the jury is still out on any definitive or final conclusion; however, many feel SSM is a very good start.

SUSTAINABLE MATERIALS MANAGEMENT

The foundation of Sustainable Materials Management (SMM) is buttressed by USEPA's recognition that no single waste management approach is suitable for managing all materials and waste streams in all circumstances. Because of this recognition, USEPA developed the non-hazardous materials and waste management hierarchy (see Figure 3.1). As shown in Figure 3.1, USEPA's hierarchy places emphasis on reducing, reusing, and recycling as the key to Sustainable Materials Management.

SOURCE REDUCTION AND REUSE

Source reduction (aka waste prevention) means reducing waste at the source and is the most environmentally preferred strategy. Properly employed source reduction can save natural resources, conserve energy, reduce pollution, reduce the toxicity of waste, and save money for consumers and businesses. Source reduction can take many different shapes or forms, including reusing or donating items, buying in bulk, reducing packaging, redesigning products, and reducing toxicity. Source reduction is also important in manufacturing. Lightweighting of packaging, whereby a reduction

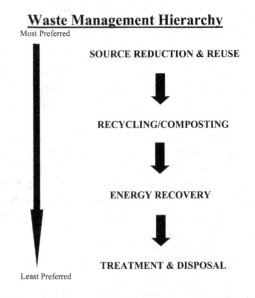

FIGURE 3.1   USEPA's waste management hierarchy. Source: USEPA (2017).

in the weight of a particular product or material per unit of material contained in the package, is becoming a more popular and cost-effective business trend. An example of this practice is producers using less plastic in food containers and less aluminum in a beverage can. Other popular and cost-effective business trends include reuse and remanufacturing. Purchasing products that incorporate these features support source reduction.

RECYCLING AND COMPOSTING

Recycling is a series of activities that includes collecting used, reused, or unused items that would otherwise be considered waste; sorting and processing the recyclable products into raw materials; and remanufacturing the recycled raw materials into new products. The benefits of recycling include preventing the emission of many greenhouse gases and water pollutions; saving energy; supplying valuable raw materials to industry; creating jobs; stimulating the development of greener technologies; conserving resources for our children's future; and reducing the need for new landfills and combustors. Note that it is the consumers who provide the last link in recycling by purchasing products made from recycled content. Recycling can also include composting of food scraps, yard trimmings, and other organic materials (USEPA, 2017).

ENERGY RECOVERY

Energy recovery from waste is the conversion of non-recyclable waste materials into usable heat, electricity, or fuel through a variety of processes, including combustion, gasification prydization, anaerobic digestion, and landfill gas (LFG) recovery. This process is often called waste-to-energy (WTE). Converting non-recyclable waste materials into electricity and heat generates a renewable energy source and reduces carbon emissions by offsetting the need for energy from fossil sources, which reduces methane generation from landfills. After energy is recovered, approximately 10% of the volume remains as ash, which is generally sent to a landfill as waste (USEPA, 2017).

DID YOU KNOW?

Waste-to-energy plants reduce 2,000 lb. of garbage to ash weighing approximately 300–600 pounds, and they reduce the volume of wastes by 87% (EIA 2020).

TREATMENT AND DISPOSAL

Treatment of waste can help reduce the volume and toxicity of waste. Treatments can be physical (e.g., shredding), chemical (e.g., incineration), and biological (e.g., anaerobic digestion). One of the most important components of an integrated waste management system is waste disposal in landfills. Modern landfills are well-engineered facilities, located, designed, operated, and monitored to ensure compliance with state and federal regulation. Landfills that accept municipal solid waste are primarily regulated by state, tribal, and local governments. EPA, however, established national standards that these landfills must meet in order to stay open. Today's landfills must meet stringent design, operation, and closure requirements. Methane gas, a byproduct of decomposing waste, can be collected and used as fuel to generate electricity. After a landfill is capped, the land may be used for recreation sites such as parks, golf courses, and ski slopes.

IS EPA'S SSM THE SOLUTION TO WASTE DISSOLUTION?

USEPA's Sustainable Management plan and its hierarchy are giant steps forward but the question remains: Is SSM the solution to waste dissolution? Again, SSM is a giant step forward but unless waste is not created, there is no zero production of waste. And as has been said throughout this presentation, waste does not go away. So the question becomes, is there a way to prevent waste or a methodology to control and eliminate waste, and if not, is there a better methodology than SSM in dealing with waste?

Yes, there is a better way to treat waste as compared to the SSM model. The problem with SSM is it is a linear approach, commonly identified as the Linear Economy Approach, to dealing with waste as shown in Figure 3.2.

The current thinking on waste management is the departure from a linear economy (as shown in Figure 3.2) to a zero waste economy. Is zero waste possible? Again, if waste is not produced in the first place, then zero waste is possible. However, not producing waste in the first place is almost impossible. This is not to say that waste can't be reduced. Waste can be reduced but not in a linear fashion. What is needed is a different approach instead of the take, make, dispose model of waste production and processing.

One approach that has been put forward is called the circular economy approach to handling waste. In this model (see Figure 3.3), the economic system is aimed at eliminating waste and the continual use of resources. In its simplest form, the circular economy is about changing the way we produce, assemble, sell, and use products to minimize waste and to reduce our environmental impact. This model is also a bonus for businesses in that it ensures maximizing the use of valuable resources by keeping equipment and infrastructure in use for long, thus improving the productivity of these resources. The bottom line in the circular economy is that waste materials and energy become input for other processes making the process regenerative in nature. This allows the goal to be reached in keeping the waste out of the garbage can.

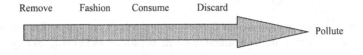

FIGURE 3.2   Linear economy dealing with waste. Source: Adaption from C. Weetman (2020).

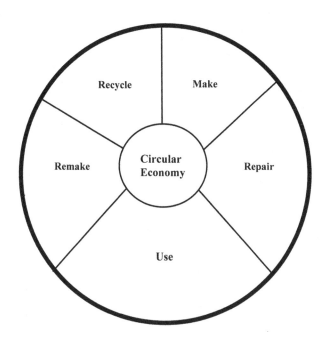

FIGURE 3.3 Circular economy approach to handling waste.

REFERENCES

Drexler, E. (1986). *Engines of Creation: The Configuration of Nanotechnology*. New York: Anchor.

EIA (U.S. Energy Information Administration). (2020). Biomass explained. Accessed 1/29/2021 @ https://www.eia.gov/energyexplaiend/biomass/wste-to-energy.php.

USEPA. (2017). Sustainable materials management: Non-hazardous materials and waste management hierarchy. Accessed 1/29/2021 @ https://www.epa.gov/smm/sustainable-materials-management-non-hazardous-materials-and-waste-management-hierarchy.

Weetman, C. (2020). *A Circular Economy*, 2nd ed. New York: Kogan Page Inc.

4 Litter

I see litter as part of a long continuum as anti-social behavior.

—**Bill Bryson**

BREAK OUT THE HOSE

Have you ever walked by or drove by or walked into or drove into one of those ice cream places, stores, shops, or creameries and noticed someone out in the parking lot with a hose washing down the lot on a daily basis? Sometimes the person with the water hose is using it a couple of times each day.

Why? Why the routine hose down of the parking lot area?

This question is best answered by checking out the ice cream establishments prior to the hose washdowns. When this is done, you might be amazed, shocked, surprised, and hopefully appalled at the amount of ice cream, chocolate fudge topping, cherries, nuts, and assorted containers and napkins and straws littering the ground, the parking lot. You also readily see the rivers of ice cream and toppings that make their way here, there, and everywhere in the lot. Ice cream and toppings are messy, sticky, runny, pasty, and downright nasty. Have you ever tried to clean up spilled or thrown down on the ground ice cream and its accompanying dressings or toppings? A real nasty clean-up mess. And this is the reason that you will always spot a hose or a hose spigot outside the ice cream shop, creamery, or store. Washing away ice cream residue is the easiest way to clean up the mess, and use the hose stream to wash away the mess. However, when washing away the mess, where does it end up? Remember, throwing away of disposing waste does not make it go away; it ends up somewhere, and that somewhere might be worse than the ice cream store parking lot.

Oh, by the way, when you pass by or go to the local ice cream parlor, store, shop, or creamery, besides observing the water spigot and hose you will also notice trash cans. Usually the trash cans are placed in convenient, easy-to-get-to locations in front of the store and near the parking areas. So the question becomes, how does the ice cream mess end up on the parking lot or parking area grounds? If you have observed someone park their vehicle, get out, and go inside the ice cream shop, order, receive their order, and then go back to their vehicle and climb inside and eat their ice cream, you might know the answer. Have you ever spotted someone who opens their car window and throws the ice cream containers and/or left-over ice cream out the vehicle window and onto the parking lot ground? Maybe you have because it is a common practice.

Why? Trash cans are placed close by so why dump the mess? Well, when I interviewed a couple of ice cream dumpers they sarcastically replied that they paid for the product and it is someone else's duty to clean up the mess.

DOI: 10.1201/9781003252665-4

Someone else's mess?

Is that not what littering is all about? It is someone else's mess.

A MIDNIGHT CAPER

In addition to ice cream, litter consists of other waste products that have been tossed out incorrectly, without permission, at an unacceptable location. We think of the term "litter" as a noun, an object improperly disposed of. But it is common practice to use the term "litter" as a verb, meaning to drop and leave objects, often human-made, such as paper cups, fast food wrappers, aluminum cans, cardboard boxes, or plastic bottles on the ground.

Cleaning up the mess is someone else's problem, someone else's job to dispose of the mess as opposed to disposing of them correctly. Right?

Then there is the so-called "midnight dumping" (aka "fly dumping" or "wildcat dumping") problem. This illegal dumping is a major problem in many communities throughout the United States. Simply, illegal dumping is the disposal of waste in an unpermitted area. Materials are often dumped from vehicles along roadsides and later at night because it saves money and hassle. It usually costs money to deposit trash and any unwanted material in a permitted dump site. Also, the legal dump sites are not always close by; it is a hassle to drive a distance to the legal dump site—it takes time and effort required for proper disposal. The materials typically included in the midnight dumping consist of:

- Demolition and construction waste such as roofing shingles, lumber, dry-wall, bricks, siding, and concrete
- Abandoned tires, auto parts, and automobiles
- "White goods" (appliances)
- Furniture
- Medical wastes
- Household trash
- Yard waste

The practice of midnight dumping is often performed because the item thrown out is not permitted or allowed to be disposed of in permitted dumps or landfills. Thus, the proper management of such wastes can be costly. If a residence or commercial activity is located in areas that lack or have costly pickup service, wastes are often illegally dumped.

When thinking about midnight dumping, it is common to think of the possible dump site or sites as being located in the woods somewhere, in a place that is secluded, out of sight. Actually, sites for midnight dumping of industrial, residential, or commercial waste can be found almost anywhere, on vacant lots, on public or private land, and also in infrequently used roadways and alleys. Poor lighting and accessibility are the two main factors in finding a spot for midnight dumping. This is the reason that areas along rural roads and railways a particularly inviting to the illegal dumper. Note that illegal dumping can occur at any time or day but is more common at night or in the early morning hours during warmer months.

One of the major problems with midnight dumping is that if the problem is not addressed, illegal dumps often attract more waste, potentially including hazardous wastes such as household chemicals and paints, automotive fluids, asbestos, and commercial or industrial wastes.

So who are the illegal dumpers?

Good question. It is very difficult to profile a "typical" illegal dumper. Offenders can include just about anyone and generally include:

- Demolition, construction, roofing, landscaping, and remodeling contractors
- Operators of junkyards or transfer stations
- Scrap collectors
- Local residents and "do-it-yourselfers"
- Waste management companies or general hauling contractors
- Automobile repair or tire shops

SALT ON THE WOUND

Not only does litter have an environmental impact; it also has a human impact. With regard to the environment, litter disposed of in the environment can be persistent, taking a long time before decomposition, and can be transported over large distances in the world's oceans. Litter can affect the quality of life; it's like putting salt on a wound and making the consequences much more painful.

Let's look at examples of tire-dumping and the creation of mosquito havens and their effects on the environment and on humans:

CASE STUDY 4.1 PROBLEM WASTES—ILLEGAL TIRE DISPOSAL

Since the invention of the automobile, what to do with worn-out tires has presented a disposal problem. America's love affair with cars means that hundreds of millions of tires are discarded every year. Stockpiles across the country store billions of worn-out tires. Some of these stockpiles are legal, others are not, but all present us with problems, including the risk of catastrophic fire and the creation of prime breeding habitat for mosquitoes, some varieties of which carry encephalitis (DEP Issue Profile, 1994).

In recent years, tire fires have received national attention. Tire fires are particularly difficult to extinguish. Water used as an extinguishing agent can cause oily run-off and burning tires emit toxic black smoke, causing pollution problems for air, surface and groundwater supplies, and soil. Smothering the fire with dirt and sand appears to be the most cost-effective and efficient way to control burning tire stockpiles. However, sometimes more unusual problems occur with tire fires.

In January 1996, a major road in Ilwaco, Washington, began to heat up, and two months later, created a major oil leak, resulting in a massive underground fire. While response teams immediately contained the oil, they were forced to allow the fire to smolder while they figured out how to even get to it (Waste Handling Equipment News, 2009).

Using scrap tires as sub-grade road base is probably the most successful effort at legally recycling used tires, along with using shredded tires as supplemental fuel for modern, scrubber-equipped boilers. However, the chances of further episodes of burning roads, which could create contamination problems in all our environmental media, means that the risks for that use of scrap tires are great. As more states enact legislation prohibiting the disposal of tires in landfills, recycling, stockpiling, and waste tire dumps will increase, until we can properly manage, store, and process these wastes (PA Dept. of Environmental Protection, 2009).

Hazardous materials encapsulated in tires can leach into water sources, contaminate the soil, and pollute the air.

Illegal burning of tires is like adding salt to the wound.

CASE STUDY 4.2 MOSQUITO HAVENS

With the virus issue affecting most of the world's population, the last thing we need is vectors like mosquitoes carrying other diseases to humans and animals. Improperly discarded items such as open containers—paper cups, cardboard food packets, plastic drink bottles, and aluminum cans get filled up with rainwater, providing breeding locations for mosquitoes. Litter can also clog drains, creating breeding havens—habitats—for mosquitoes. Mosquitoes can spread several diseases like malaria, dengue, West Nile virus, and others.

Providing habitat for mosquitoes is like adding salt to the wound.

REFERENCES

DEP Issue Profile. (1994). Scrap tire management. Accessed 02/02/2021 @ http://www.state.me.us.

PA. Dept. of Environmental Protection. (2009). Tire dumps: Risks and dangers. Accessed 02/04/2021 @ http://www.efp.state.pa.us.

Waste Handling Equipment News. (2009). Canadian hi tech firm finds used tire solution. Accessed 02/02/2021 @ http://www.ewmc.com.

5 Ocean Dumping

Water and air, the two essential fluids on which all life depends, have become global garbage cans.

—Jacques Cousteau

The ocean and the islands and beaches that once were so fair, that
humans could find a sandy stretch and relax without much care.
Then came the remnants of waste cast away by those who didn't
understand, how to protect the lakes, rivers, seas and sacred land.
They built ramps, dumps, boats to cast away their trash that affect fish,
and then people use nets, hooks, lines, and sinkers to fill their dish.
From their dish full of fish became the order of the day and their fill,
this made sick the people and the rush to fill their wills.
To bring the people and the oceans back to health, from dire
consequences, humans must avoid their stealth.

—Frank R. Spellman (1996)

In 1980, as an officer in the U.S. Navy I was stationed aboard an amphibious warship that was underway from Norfolk, Virginia, to Rota, Spain. We navigated the trip via the Great Circle Route which is the shortest route between two points on the surface of a sphere. After we got underway and were two days on route to Rota, I served my routine four-hour watch as Officer of the Deck (OOD), navigating the ship along its charted track. The ship was equipped with navigation hardware including gyrocompass, magnetic compass, and satellite navigation (SATNA for geospatial positioning) aids. And the duty Quartermaster was diligently plotting our track on the map.

During my OOD watches, I quickly concluded that if all navigation aids available to me had failed I would have no problem keeping the ship on the proper course. All I needed to do was to ensure that the helmsmen paralleled the stream of garbage that was apparent and continuous on our port side during most of the trip.

OCEAN DUMPING IS NOTHING NEW

A couple of years ago I performed a non-scientific survey of more than 60 people from all walks of life on ocean dumping. The purpose of the survey? I wanted to get an idea of what people thought of the practice of dumping trash and other unwanted items into the oceans. Actually, from the responses I was able to obtain, I ended up

DOI: 10.1201/9781003252665-5

with more information than I expected. For example, I was surprised to find out that many of the respondents not only thought dumping trash in the oceans was "no big thing" but also that the practice was okay and is currently being conducted universally without regulation.

I guess I should not have been all that surprised by the responses. This is the case, of course, because in the past (before 1972), communities around the world used the ocean for waste disposal, including the disposal of chemical and industrial wastes, radioactive wastes, trash, munitions, sewage sludge, and contaminated dredged material. Little to no attention and/or concern was given or expressed to the negative impacts of waste disposal on the marine environment. Even less attention was focused on opportunities to recycle or reuse such materials.

While reviewing many of the comments I obtained from respondents to my questions on ocean dumping, I noticed a trend of thinking or mindset similar to the 19th-century frontier mentality that settled the wide-open western United States. I noticed that the mindset that humans are superior to all else on Earth and that the oceans are endless was prevalent. And so why not use them to dump what is no longer needed or wanted? Is this an "out of sight, out of mind" idiom at work? Partially this is the case, but again it is the mindset that the oceans are endless just as the Old West was in the United States, and dumping unwanted materials in one location will have zero effect on the next location. With regard to the oceans, wastes were frequently dumped in coastal and ocean waters based on the assumption that marine waters had an unlimited capacity to mix and disperse waste. This is the idiom that says, "Dilution is the solution to pollution." We experience this dilution is the solution to pollution idiom every day whenever a smokestack sends dirty smoke bellowing out into the atmosphere, and the air above the stack is so immense that with time and wind, it dilutes the smoke, and it is no longer apparent. The problem with dilution is the solution to pollution mindset or idiom is that it is totally dependent on two things: frequency of dumping and quantity of pollution. Consider, for example, Figure 5.1, where end-of-pipe pollution is shown dumping into a river body—remember, all rivers empty into the sea or ocean. If this particular end-of-pipe pollution is the only source of pollution into this particular river, then the flow and turbulence within the river, if its total stretch is long enough, will dilute the pollution and the downstream water will return to normal—depending on what is normal, of course.

To gain a better understanding and appreciation of the fallacy portended by the "dilution is the solution to pollution" idiom is shown in Figures 5.2 and 5.3. The point is that dilution can be the solution only when there are no multiple sources of pollution dumped in close proximity with each other. As shown in the urban water cycle, rivers usually are the locations of not one city or town but more than one. When this is the case, dilution is the solution to pollution is not possible because dilution can't occur when pollutants are summarily dumped downstream one after another. In the indirect water reuse process, the same issue is apparent. A river with one city or town using its water and discharging used water into it is workable. However, when the river is home to more than one settlement, dilution may not be the solution to pollution.

FIGURE 5.1 End-of-pipe pollution. Source: Illustration by Kat Welsh-Ware and F. R. Spellman.

We can only guess or estimate the volumes and types of materials disposed of in ocean wastes in the United States prior to 1972 because no complete records exist. However, some reports indicated that vast amounts of waste were ocean dumped. For example, in 1968, the National Academy of Sciences estimated annual volumes of ocean dumping by vessels or pipes:

- 100 million tons of petroleum products
- 2–4 million tons of acid chemical wastes from pulp mills
- More than 1 million tons of heavy metals in industrial wastes
- More than 100,000 tons of organic chemical wastes

A 1970 Report to the President from the Council on Environmental Quality on ocean dumping described that in 1968, the following were dumped in the ocean in the United States:

- 38 million tons of dredged material (34% of which was polluted)
- 4.5 million tons of industrial waste
- 4.5 million tons of sewage sludge (aka biosolids—significantly contaminated with heavy metals)
- 0.5 million tons of construction and demolition debris

EPA records indicate that more than 55,000 containers of radioactive wastes were dumped at three ocean sites in the Pacific Ocean between 1946 and 1970. Almost

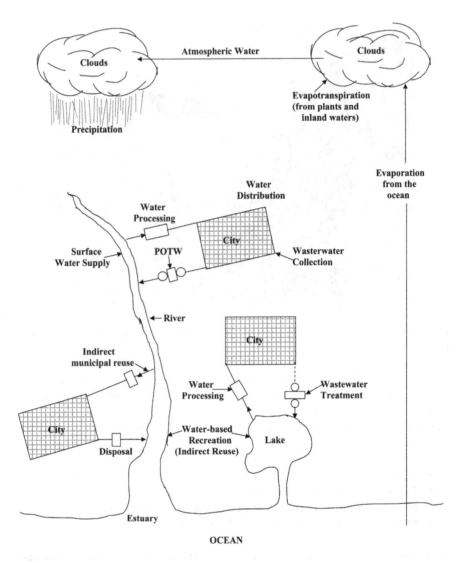

FIGURE 5.2 Urban water cycle.

34,000 containers of radioactive wastes were dumped at three ocean sites off the East Coast of the United States from 1951 to 1962.

Following decades of uncontrolled dumping, some areas of the ocean became demonstrably contaminated with high concentrations of harmful pollutants including heavy metals, inorganic nutrients, and chlorinated petrochemicals. The uncontrolled ocean dumping caused severe depletion of oxygen levels in some ocean waters. At the mouth of the Hudson River (aka the New York Bight) where New York City dumped sewage sludge and other materials, oxygen concentrations in waters near the seafloor declined significantly between 1949 and 1969.

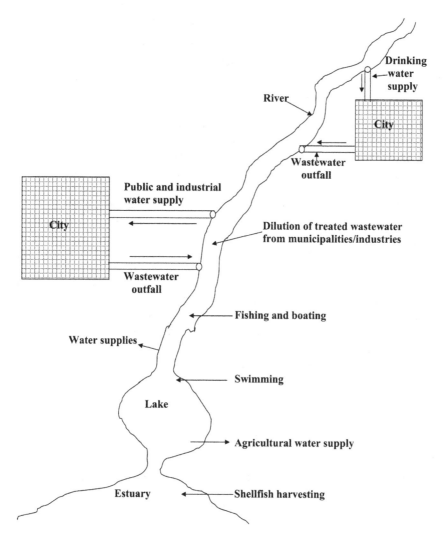

FIGURE 5.3 Indirect water reuse process.

DID YOU KNOW?

We all understand that air contains oxygen but few of us realize that water also contains small amount of oxygen. Oxygen in water is necessary to maintain aquatic life. When the same water is overloaded with organics in the presence of oxygen, the oxygen is used to biodegrade the organic material. Too much organic material in water means oxygen is depleted to the point where it is used up and the dissolved oxygen decreases to the point where aquatic life and the quality of the water is affected.

MARINE PROTECTION, RESEARCH, AND SANCTUARIES ACT

Congress enacted the Marine Protection, Research, and Sanctuaries Act (MPRSA) in 1972 (aka Ocean Dumping Act). The Act declares that it is the policy of the United States to regulate the dumping of all materials which would adversely affect human health, welfare or amenities, the marine environment, ecological systems, or economic potentialities.

What the MPRSA does is it implements the requirements of the London Convention (Convention on the Prevention of Marine Pollution by Dumping Wastes and Other Matter of 1972). The London Convention is one of the first international agreements for the protection of the marine environment from human activities. The MPRSA consists of Title 1 which contains the permitting and enforcement provisions for regulating ocean dumping and Title II authorizes marine research.

USEPA uses MPRSA to establish criteria for reviewing and evaluating permit applications. USEPA is responsible for issuing ocean dumping permits for materials other than dredged material. With regard to dredged material, the U.S. Army Corps of Engineers (USACE) is responsible for issuing ocean dumping permits, using USEPA's environmental criteria. Permits for ocean dumping of dredged material are subject to EPA review and written concurrence. EPA is also responsible for designating and managing ocean disposal sites for all types of materials (USEPA, 2020).

The site management and monitoring plans (SMMPs) for each designated ocean dredge materials disposal site are co-developed by EPA and USACE. Moreover, EPA and USACE often work together in conducting oceanographic surveys at ocean disposal sites to evaluate environmental conditions at the site and to determine what management actions may be needed.

DID YOU KNOW?

40 CFR 220–229 contains EPA's ocean dumping regulations and includes the criteria and procedures for ocean dumping permits and for the designation and management of ocean disposal sites under the MPRSA. USACE has published regulations under various provisions of 33 CFR 320, 322, 324, 325, 329, 331, and 335–337.

Environmental regulations employed by EPA and USACE are important because they prevent unregulated disposal of wastes and other materials into the ocean which results in the degradation of marine and natural resources and poses human health risks. For almost 50 years, EPA's Ocean Dumping Management Program has prevented many harmful materials from being ocean dumped, worked to limit ocean dumping generally, and worked to prevent adverse impacts to human health, the marine environment, and other legitimate uses of the ocean (e.g., fishing and navigation) (USEPA, 2020).

Note that EPA under the MPRSA is not the only federal agency that has a role with respect to ocean dumping. The USACE, the National Oceanic and Atmospheric Administration (NOAA), and the U.S. Coast Guard (USCG) also have roles with respect to ocean dumping. While it is true that USEPA has primary authority for regulating ocean disposal of all materials, it does not have full responsibility for the regulation of dredged material disposal in the ocean. This responsibility is shared with USACE. Maintaining surveillance of ocean dumping is the responsibility of USCG. In conducting long-range research on the effects of human-induced changes to the marine environment, is NOAA's responsibility.

In addition, USEPA's Ocean Dumping Management Program coordinates with partners at the international, federal, state, and local levels, and through inter-agency groups, including National and Regional Dredging Teams, on ocean dumping, dredged material management, pollution prevention, and marine protection activities.

WHAT'S BEING OCEAN DUMPED TODAY?

At the current time, the vast majority of material disposed of in the ocean is uncontaminated dredged material (sediment) removed from our nation's waterways to support a network of coastal ports and harbors for commercial, transportation, national defense, and recreational purposes. Additional materials disposed of in the ocean include human remains for burial at sea, vessels, human-made ice piers in Antarctica, and fish wastes.

WHAT CANNOT BE DUMPED IN THE OCEAN?

USEPA and the MPRSA ocean dumping regulations prohibit ocean dumping of certain materials, such as:

- Radiological, chemical, and biological warfare agents
- High-level radioactive wastes
- Persistent inert synthetic or natural materials which may float or remain in suspension in the ocean in such manner that they may interfere materially with fishing, navigation, or other legitimate uses of the ocean
- Sewage sludge
- Materials insufficiently described to permit application of the environmental impact criteria of 40 CFR 227 subpart B
- Medical wastes (isolation wastes, infectious agents, human blood and blood products, pathological wastes, sharps, body parts, contaminated bedding, surgical wastes and potentially contaminated laboratory wastes, and dialysis wastes)
- Industrial wastes, specifically liquid, solid, or semi-solid wastes from a manufacturing or processing plant (except on an emergency basis)
- Materials containing the following constituents in greater than trace amounts (except on an emergency basis): organohalogen compounds, mercury, and mercury compounds
- Cadmium and cadmium compounds, oil of any kind or in any form, and known carcinogens, mutagens, or teratogens

DID YOU KNOW?

Ocean dumping is usually placed into three lists: gray list (highly contaminated and toxic water); blacklist (mercy, cadmium, plastic, oil products, radioactive waste, and anything made for biological and chemical warfare); whitelist contains every other material not already mentioned in the gray and blacklist and those materials that disturb or damage the coral reef ecosystems.

OCEAN DUMPING BAN ACT

The 1988 Ocean Dumping Ban Act banned the dumping of industrial wastes, including those previously permitted for incineration at sea. Incineration at sea is considered to be ocean dumping because of the emissions from the stack deposit into the surrounding ocean waters. Moreover, incineration at sea is regulated under the London Convention and London Protocol. The London Convention defines "incineration at sea" as the deliberated combustion of wastes and other matter in marine incineration facilities for the purpose of their thermal destruction. Combustion associated with activities incidental to the normal operation of vessels, platforms, and other man-made structures is excluded from the scope of this definition. Note that marine incineration facility means a vessel, platform, or other man-made structure operating for the purpose of incineration at sea.

Under the London Protocol, incineration at sea and the export of wastes and other materials for incineration at sea are prohibited. The United States has signed the London Protocol, which is intended to modernize and eventually replace the London Convention; however, the Senate has not ratified the treaty (USEPA, 2020).

IS THE 1972 MPRSA EFFECTIVE?

The ocean is no longer considered an appropriate disposal location for most wastes. The driving force for this change was the passage of the MPRSA in 1972; it is the major ruling for turning concern toward protecting the marine environment and has been quite effective. Today, the United States is at the forefront of protecting coastal and ocean waters from adverse impacts due to ocean dumping.

CASE STUDY 5.1 RIVER CLEANUP

Volunteers for the biannual Conestoga River cleanup found everything from tires, washing machines, bicycles, and motor scooters to candy wrappers and car engines in the river—several truckloads of trash. Low water levels enabled volunteers to remove trash in the middle of the river, where in higher water seasons, it has been unreachable. They will sell the metal for scrap, and the

Lancaster County Solid Waste Authority will waive the tipping fee for disposing of the rest of the illegally dumped trash.

Volunteers also planted bushes (donated by the Chesapeake Bay Foundation) on the stream banks, to serve the dual function of erosion and dumping prevention. Organizers see signs that the river cleanup program is helping in several ways. More volunteers turn out each time for the cleanup, allowing the group to expand their coverage. They also feel the message is getting through to people that the river is not the place to dump their trash. The river gives up less trash each year. State Representative Mike Sturla hopes that people are beginning to understand that "this is not the place to put it" (Lancaster New Era, 9/27/98).

DID YOU KNOW?

It is estimated that about 80% of marine debris originates as land-based trash and the remaining 20% is attributed to at-sea international or accidental disposal or loss of goods and waste. This breakdown can vary depending on factors like the location of landmasses, population densities, and behavior of currents in surrounding marine waters (USEPA, 2017).

AND THE SEA WILL FLOAT

A SEA OF PLASTIC

The sea of plastic has often been referred to as *plastic soup*. More appropriately it might be referred to as synthetic polymer soup. The term was coined by Charles J. Moore in 1997 after he found patches of plastic pollution in the North Pacific Gyre called the *Great Pacific Garbage Patch* (Plastic Soup Foundation, 2019). Plastic is distributed by currents and winds and is most

abundant in the central and western North Pacific (Day et al., 1988). Plastics have now made themselves a permanent part of the marine environment for the first time in the extended history of global seas—no ancient deposits of these materials or their biological consequences with high concentrations of synthetic polymers in the globe's one vast prehistoric ocean. However, if we look today at ice and sediment core samples, we will find a clear record of plastic deposition and the consequences (Moore, 2012).

The sea of plastic by any other name is an accumulation of plastic objects and particles (e.g., plastic bottles, bags, and microbeads) in the Earth's environment that adversely affects wildlife, wildlife habitat, and humans (Parker, L., 2018). Plastic marine debris is of particular concern due to its longevity in the marine environment, the physical (e.g., entanglement, gastrointestinal blockage, and reef destruction) and chemical threats and hazards (e.g., bioaccumulation of the chemical ingredients of plastic or toxic chemicals sorbed to plastics) it presents to marine and bird life, and the fact that it is frequently mistaken as food by birds and fish. The most attention paid to plastics contamination of marine environments is focused on the so-called "Great Pacific and Atlantic Garbage Patches," located in remote gyres (i.e., giant circular oceanic surface currents); however, it is important to note that the gyre accumulations are not the only water bodies polluted by plastics. For example, plastic trash and particles are now found in most marine and terrestrial habitats, including the deep sea, Great Lakes, coral reefs, beaches, rivers, and estuaries.

A significant amount of marine plastic waste normally falls within the broad category of macroscopic pollutants. Macroscopic pollutants include large visible items (e.g., floatable, flotsam and jetsam, nurdles, marine debris, and shipwrecks); these contaminate or pollute surface water bodies (lakes, rivers, streams, and oceans). In an urban stormwater context, these large visible items are termed *floatable*— waterborne litter and debris, including toilet paper, condoms, tampon applicators, plastic bags or six-packs rings, and accompanied trash such as food cans, jugs, cigarette butts, yard waste, polystyrene foam, and metal and glass beverage bottles (see Figures 5.4–5.6), as well as oil and grease. Floatable come from street litter that ends up in storm drains (catch basins) and sewers. Floatable can be discharged into the surrounding waters during certain storm events when water flow into treatment plants (i.e., those without overflow storage lagoons) exceeds treatment capacity. Floatable contribute to visual pollution, detract from the pleasure of outdoor experiences, and pose a threat to wildlife and human health.

FIGURE 5.4 Floatable litter in beach grass along shore of the Chesapeake Bay, Norfolk, Virginia. Source: Photo by F. R. Spellman.

FIGURE 5.5 Floatable plastic bottle at Ocean View Beach, Chesapeake Bay, Norfolk, Virginia. Source: Photo by F. R. Spellman.

FIGURE 5.6 Nurdles. Source: Photo by F. R. Spellman.

DID YOU KNOW?

Almost every type of commercial plastic is present in aquatic/marine debris. However, it is interesting to note that the floating compounds are dominated by polyethylene and polypropylene because of their high production volumes, their broad utility, and their buoyancy (USEPA, 2017).

The terms *flotsam* and *jetsam*, as used currently, refer to any kind of marine debris. The two terms have different meanings: jetsam refers to materials jettisoned voluntarily (usually to lighten the load in an emergency) into the sea by the crew of a ship, and flotsam describes goods that are floating on the water without having been thrown in deliberately, often after a shipwreck.

A *nurdle* (strongly resembling a fish egg), also known as mermaids' tears, pre-production plastic pellet, or plastic resin pellet, is a plastic pellet typically under 5 mm in diameter (see Figure 5.6) and is a major component of marine debris. It is estimated that at least 60 billion pounds of nurdles are manufactured annually in the United States alone (Heal the Bay 2009). Not only are they a significant source of ocean and beach pollution; nurdles also frequently find their way into the digestive tracts of various marine creatures.

In the past, the major source of *marine debris* was naturally occurring driftwood; humans have been discharging similar material into the oceans for thousands of years. In the modern era, however, the increasing use of plastic with its subsequent discharge by humans into waterways has resulted in plastic materials and/or products being the most prevalent form (as much 80%) of marine debris. Plastics are persistent water pollutants because they do not biodegrade as many other substances do. Not only is waterborne plastic unsightly; it poses a serious threat to fish, seabirds, marine reptiles, and marine mammals, as well as to boats and coastal habitations (NOAA 2009). Plastic debris has been responsible for the entanglement and deaths of many marine organisms, such as fish, seals, turtles, and birds. Sea turtles are affected by plastic pollution. Some species consume jelly fish but often mistake plastic bags for their natural prey. Consumed plastic materials can kill the seat turtle by obstructing its esophagus (Gregory, 2009). Baby sea turtles are particularly vulnerable according to a 2018 study by Australian scientists (Gabbatiss, 2018).

DID YOU KNOW?

It has been estimated that container ships lose over 10,000 containers at sea each year (usually during a storm; Podsada 2001). One famous spillage occurred in the Pacific Ocean in 1992, when thousands of rubber ducks and other toys went overboard during a storm. The toys have since been found all over the world; scientists have used the incident to gain a better understanding of ocean currents.

TYPES OF PLASTIC

Types of plastics include:

- Polyethylene terephthalate (PET or PETE)—the largest use of PET is for synthetic fibers, in which case it is referred to as polyester. PET's next largest application is as bottles for beverages, including water. It is also used in electrical applications and packaging (ICS, 2011a).
- High-density polyethylene (HDPE)—HDPE is used for a wide variety of products, including bottles, packaging containers, drums, automobile fuel tanks, toys, and household goods. It is also used for packaging many household and industrial chemicals such as detergents and bleach and can be added to objects such as crates, pallets, or packaging containers (ICS, 2011b).
- Polyvinyl chloride (PVC)—PVC is produced as both rigid and flexible resins. Rigid PVC is used for pipe, conduit, and roofing tiles, whereas flexible PVC has applications in wire and cable coating, flooring, coated fabrics, and shower curtains (ICIS, 2011c).
- Low-density polyethylene (LDPE)—LDPE is used mainly for film applications in packaging, such as poultry wrapping, and in on-packaging, such as trash bags. It is also used in cable sheathing and injection molding (ICIS, 2011b).
- Polypropylene (PP)—PP is used in packaging and automotive parts, or made into synthetic fibers. It can be extruded for use in pipes, conduits, wires, and cables. PP's advantages are high-impact strength, high softening point, low density, and resistance to scratching and stress cracking. A drawback is its brittleness at low temperatures (ICIS, 2011d).
- Polystyrene (PS)—PS has applications in a range of products, primarily domestic appliances, construction, electronics, toys, and food packaging such as containers, produce baskets, and fast-food containers (ICIS, 2011e).
- Thermosets—Thermosets are plastics and resins that harden permanently after one treatment of heat and pressure; once they are set, they can't be melted.
- Thermoplastics—A thermoplastic is a type of plastic that melts when heated and freezes to a solid when cooled sufficiently.
- Elastomers—Elastomers are polymers with viscoelasticity (i.e., both viscosity and elasticity) and have very weak intermolecular forces, generally low Young' modulus (elasticity in tension) and high failure strain compared with other materials (De, 1996).

DID YOU KNOW?

The primary contributors of plastics in the oceans are China, Indonesia, the Philippines, Sri Lanka, and Thailand (Spellman, 2014).

MICROPLASTICS

Approximately 90% of the plastics in the pelagic marine environment are less than 5 mm in diameter; these are known as microplastics (Eriksen et al., 2013; Browne et al., 2010; Thompson et al., 2004). Microplastics are generated when larger pieces of plastic are fragmented by weathering actions including the effects of ultraviolet rays and wind and wave action. Current information on the use of tiny plastic abrasives (commonly called microbeads or nanobeads), especially in pharmaceuticals and personal care products (PPCPs) and home cleaning products, and synthetic fabrics shedding during laundering has shown the prevalence of micro- and nanoparticle-size plastics as being pervasive in water bodies (Eriksen et al., 2013). With regard to PPCPs, they are sometimes termed "emerging pollutants," it is important to point out that PPCPs are not truly emerging; it is the understanding of the significance of their occurrence in the environment that is beginning to develop. PPCPs comprise a very broad, diverse collection of thousands of chemical substances, including prescription, veterinary, over the counter (OTC) therapeutic drugs, fragrances, cosmetics, sunscreen agents, diagnostic agents, nutraceuticals (vitamins), biopharmaceuticals (medical drugs produced by biotechnology), growth-enhancing chemicals used in livestock operations, and many others. Again, plastics are used in some of these products mainly in cosmetics. This broad collection of substances refers, in general, to any product used by individuals for cosmetic reasons (e.g., anti-aging cleansers, toners, exfoliators, facial masks, serums, and lip balm).

Note that wastewater treatment processes remove many substances and particles from the waste stream; however, this may not be the case with PPCPs and plastics—they largely pass through unchanged. The micro- and nanoparticle plastics, as well as other microplastics caused by fragmentation, are available for ingestion by a wide range of animals in the aquatic food web. The problem with PPCPs is that we do not know what we do not know about them—the jury is still out on their exact environmental impact.

Even though the jury is still out on PPCPs and although PPCPs are used in large quantities, the concentrations of PPCPs currently being found in water suppliers are very small. The laboratory tests for these compounds do not report concentrations in parts per million (ppm) or parts per billion (ppb) but instead report concentrations in parts per trillion (ppt), which is the same as nanograms per liter. One part per million is equivalent to a shot glass full of water dipped from Olympic swimming that is 2 m deep. One part per billion is equivalent to one drop from an eyedropper filled from the same Olympic pool. One part per trillion is equivalent to 1 drop in 20 Olympic pools that are 2 m deep or 1 second in 31,700 years (Spellman, 2014).

It is interesting to note that although nearly every type of commercial plastic is present in marine debris, floating marine debris is dominated by polyethylene and polypropylene because of their high production volumes, their broad utility, and their buoyancy (Colton et al. 1974; Ng and Obbard, 2006; Rios et al., 2007). Low-density polyethylene or linear low-density polyethylene is commonly used to make plastic bags or six-pack rings; polypropylene is commonly used to make reusable food containers or beverage bottle caps (USEPA, 2017).

Even though much of the marine debris research focuses on floating plastic debris, it is important to recognize that only approximately half of all plastic is definitely buoyant, that is, it floats (USEPA, 2017). Buoyance is dependent on the density of the material and the presence of entrapped air (Andrady, 2011). Researchers have documented the presence of plastics throughout the water column, including on the seafloor of nearly every ocean and sea (Ballent et al., 2013, Maximenko et al., 2012).

PLASTIC CONTAINERS AND PACKAGING

Plastic resins are used in the manufacture of packing products. Some of these include polyethylene terephthalate (PET) soft drink and water bottles, high-density polyethylene (HDPE) milk and water jugs, film products (including bags and sacks) made of low-density polyethylene (LDPE), and other containers and packaging (including clamshells, trays, caps, lids, egg cartons, loose fill, produce baskets, coatings, and closures) made up of polyvinyl chloride (PVC), polystyrene (PS), polypropylene (PP), and other resins. USEPA used data on resin sales from the American Chemistry Council to estimate the generation of plastic containers and packaging in 2017.

USEPA estimated approximately 14.5 million tons of plastic containers and packaging were generated in 2017, approximately 5.0% of municipal solid waste (MSW) generation. (Note that plastic packaging as a category in this analysis does not include single-service plates and cups and trash bags, both of which are classified as nondurable goods.)

USEPA also estimated the recycling of plastic products based on data published annually by the American Chemistry Council, as well as additional industry data. The recycling rate of PET bottles and jars was 29.1 in 2017 (910,000 tons). It is estimated that the recycling of HDPE natural bottles (e.g., milk and water bottles) was 220,000 tons, or 29.3% of generation. Overall, the amount of recycled plastic containers and packaging in 2017 was almost 2 million tons, or 13.6% of plastic containers and packaging generated. Additionally, 16.9% of plastic containers and packaging waste generated was combusted with energy recovery, while the remainder (over 69%) was landfilled (USEPA, 2021).

PLASTICS AND PERSISTENT, BIOACCUMULATIVE, AND TOXIC SUBSTANCES

Substances such as persistent, bioaccumulative, and toxic (PBTs) chemicals pose a risk to the marine environment because they resist degradation, persisting for years or even decades (USEPA, 2017). PBTs are toxic to humans and marine organisms and have been shown to accumulate at various trophic levels (feeding levels) through the food chain. PBTs can be insidious in the environment even at low levels attributable to their ability to biomagnify up the food web, leading to toxic effects at higher trophic levels even though ambient concentrations are well below toxic thresholds. The subgroup of PBTs known as persistent organic pollutants (POPs) are especially persistent, bioaccumulative, and toxic (such as DDT, dioxins, and polychlorinated biphenyls (PCBs)) (Engler, 2012).

Global representatives from 92 countries concluded that there is a definite need to reduce and eliminate worldwide use and emissions of POPs, highly toxic chemicals such as DDT, and dioxins that remain in the environment for years. The conclusion they came to was the result of agreements made during a meeting, held in Montreal, June 29 to July 3, 1998, that focused on a list of 12 persistent chemicals, including 9 pesticides. Eight of these nine pesticides are on Pesticide Action Network's Dirty Dozen list: aldrin, chlordane, DDT, dieldrin, endrin, heptachlor, hexachlorobenzene, and toxaphen. The remaining chemicals on the list are dioxins, furans, mirex, and PCBs.

"These substances travel readily across international borders to even the most remote region, making this a global problem that requires a global solution," said Klaus Toepfer, executive director of the United Nations Environment Program (UNEP), which sponsored the meetings. A growing body of scientific evidence indicates that exposure to very low doses of certain POPs can lead to cancer, damage to the central and peripheral nervous systems, diseases of the immune system, reproductive disorders, and interference with normal infant and child development. POPs can travel through the atmosphere thousands of miles from their source. In addition, these substances concentrate in living organisms and are found in people and animals worldwide.

In describing PBTs, it can be said that they are hydrophobic, aquaphobic, or have low water solubility, and for this reason, when in the marine environment, they tend to partition to sediment or concentrate at the sea surface (Hardy et al., 1990; USEPA, 2017) and not dissolve into solution. When PBTs encounter plastic debris, they tend to preferentially take up or hold (sorb) to the debris. In effect, plastics are like magnets for PBTs.

Research has shown that different pollutants sorb to different types of plastics in varying concentrations depending on the concentration of the PBT in sea water and the amount of plastic particle surface area available. Plastics on the seafloor may sorb PBTs from the sediments (Graham, Thompson, 2009; Rios et al., 2007), in addition to sorbing them from the seawater. Graham and Thompson (2009) found, like other researchers, that weathering of plastic bottles, bags, fishing lien, and other products in the ocean causes tiny fragments to break off. These plastic fragments may accumulate biofilms and sink and become mixed with sediment, where benthic invertebrates may encounter and ingest them. The concentration of PBTs such as PCBs and DDE (the breakdown product of DDT) on plastic particles has been shown to be in orders of magnitude greater than concentrations of the same PBTs found in the surrounding water.

Because of their varying behavior in the environment, the potential for PBTs to sorb to plastic debris is complex; however, they are more likely than not to preferentially sorb to plastic debris. The particular affinity to sorb will depend on the PBT and type of plastic: polyethylene sorbs PCBs more readily than polypropylene does (Endo et al., 2005). The longer plastic is in the water, the more weathered and fragmented it becomes (Teuten et al., 2007). With increased fragmentation comes higher relative surface area, thereby increasing the relative concentration of sorbed PBTs (a process referred to as hyperconcentration of contaminants) (Engler, 2012).

OTHER ASSORTED CONTAINERS AND PACKAGING

USEPA defines containers and packaging as products that are assumed to be discarded the same year the products they contain are purchased. Unfortunately, many of these containers and packaging end up in the ocean. In addition, containers and packaging make up a major portion of MSW, amounting to more than 82 million tons of generation in 2017 (28% of total generation). Packaging is the product used to wrap or protect goods, including food, beverages, medications, and cosmetic products. Containers and packaging are used in the shipping, storage, and protection of products. They also provide sales and marketing benefits.

With regard to recycling containers and packaging, it is the corrugated boxes that are among the most frequently recycled products. In 2017, the recycling rate of generated packaging and containers was 53%. Additionally, the combustion of containers and packaging was 7 million tons (21% of total combustion with energy recovery) and landfills received 30 million tons (20.5% of total landfilling) in 2017.

Containers and packaging products in MSW are made of several materials: paper and paperboard, glass, steel, aluminum, plastics, wood, and small amounts of other materials (USEPA, 2021).

GLASS CONTAINERS AND PACKAGING

When we point out that various glass container and packaging materials are a significant part of the almost endless list of waste materials dumped into the oceans, we are not referring pointedly and specifically to a "message in a glass bottle" (see Figure 5.7) pollution. A message in a bottle has been a form of communication

FIGURE 5.7 Message in a glass bottle. Source: Photo by F. R. Spellman.

dating back to early historical times. This is not to say that a message in a bottle is not pollution because it is; the growing awareness that bottles constitute waste that can harm the environment and marine life has triggered the trend in favor of using biodegradable drift cases and wooden blocks instead of glass bottles carrying messages, of any type or form (see Figure 5.7).

Glass containers typically found as ocean waste include beer and soft drink bottles, wine and liquor bottles, as well as bottles and jars for food and juices, cosmetics, and other products. Beginning in 2009, the Glass Packaging Institute provided production data, and various regulatory agencies, such as USEPA, and other interested parties use the information as needed. About 70% of glass consumption is used for containers and packaging purposes (Pongracz, 2007). From this information, USEPA estimated that 9.1 million tons of glass containers were generated in 2015, or 3.5% of MSW. At least 13.2% of the production of glass and containers are burned with energy recovery. The amount of glass containers and packaging that go into the landfill is about 53% (USEPA, 2021).

Steel Containers and Packaging

It is relatively rare to find or run into steel containers and packaging in the oceans. This is the case because only about 5% of steel is used for packaging purposes; this is in sharp contrast to the total world of steel consumption. The world production of steel containers and packaging amounted to 2.2 million tons in 2018, 0.8% of total MSW generation. Most of this amount was represented by cans used for food products. Approximately 1.6 million tons (73%) of steel packaging were recycled. Also, the steel packaging that was combusted with energy recovery was about 5%, and 21% was landfilled (USEPA 2021).

Paper and Paperboard Containers and Packaging

"Go float your boat." This is something many of us have heard as children and maybe later as adults. For those not familiar with this saying, it refers to the taking of a single piece of paper and folding it in such a fashion so as to make it "sea or bathtub worthy," so to speak. The reference here is to the paper used. Ocean waste consists of paper and cardboard containers and corrugated boxes and other paper and cardboard packaging and not handmade paper boats.

Corrugated boxes were the largest single product category of MSW in 2018 at 33 million tons generation, or 11% of total generation. Corrugated boxes also represent the largest single product of recycled paper and paperboard containers and packaging. In 2018, approximately 32 million tons of corrugated boxes were recycled out of 34 million tons of total paper and paper board recycling. The recycling rate for corrugated boxes was 230,000 tons, and landfills received 940,000 tons in 2018 (USEPA, 2021).

Other paper and paperboard packaging found in water bodies including the marine environment include milk and juice cartons and other products packaged in gable-top cartons and liquid food aseptic cartons, fold cartons (e.g., cereal boxes frozen food boxes, some department store boxes, bags, sacks, wrapping papers, and other paper and paperboard packaging (primarily set-up boxes such as, cosmetic, and candy boxes)). Overall, paper and paperboard containers and packaging totaled to about 42 million tons of MSW generation in 2018, or 14% of total generation (USEPA, 2021).

Recycling of corrugated boxes is by far the largest component of paper packaging recycling. Additionally, smaller amounts of other paper packaging products also enter the recycling stream (estimated at about 1.8 million tons in 2018). The overall recycling rate for paper and paperboard packaging was 81% in 2018. Smaller proportions were combusted for energy recovery (3.7%) and landfill (15.4%). Other paper packaging, such as sacks and cartons, is mostly recycled as mixed papers (USEPA, 2021).

Back to floating your boat. If you are into floating handmade boats composed only of wood and paper or cloth, this simple float device is made of materials that are fully biodegradable and is therefore not considered a waste product or a contaminator of marine environments.

WOOD CONTAINERS AND PACKAGING

Wood packaging includes anything made out of wood, mostly wood pallets, wood chips, boards, and planks, including wood crates. USEPA used the data of production of wood packaging derived from purchased market research and research from the Center for Forest Products at Virginia Tech and the United States Department of Agriculture' Forest Service Southern Research Station. In 2018, the data collected indicated that the amount of generated wood pallets and other wood packaging was 11.5 million tons, totaling 4% of total MSW generation. The amount of wood pallet recycling (usually by chipping for uses such as mulch or bedding material, but excluding wood combusted as fuel) in 2018 was about 3 million tons (USEPA, 2021). Additionally, about 14% of the wood containers and packaging waste generated was combusted with energy recovery, while the remaining (59%) was landfilled.

MISCELLANEOUS PACKAGING

Other miscellaneous packaging includes bags made of textiles (rags, clothing, sheets, blankets, etc.) and small amounts of leather. Estimates about the dumping of these materials are the only data currently available; however, UESPA estimated that the amount generated was 340,000 tons in 2018, and none of which was recycled or composted. Approximately 21% of the other miscellaneous packaging was combusted with energy recovery, while the majority was landfilled (79%) (USEPA, 2021).

THE BOTTOM LINE ON OCEAN DUMPING

Simply throwing waste into the ocean, whether intentional or otherwise, is performing the act of ocean dumping. To fully understand the danger of this occurrence, which is not to be taken lightly, it is necessary to know the impacts of ocean dumping. These impacts include:

- Filthy beaches
- Toxic water
- Reduced dissolved oxygen in the water
- Malnutrition in marine creatures
- Entangled marine creatures
- Poisonous sea life
- Human health problems
- Damaged sea creature internal organs
- Continued growth of ocean garbage patches
- Proliferation of invasive specifies
- Increased dead zones
- Increased death rate of sea birds
- Hindered photosynthesis for marine plants
- Disrupted food chain
- Damaged coral reef

REFERENCES

Andrady, L. (2011). Microplastics in the marine environment. *Mar. Poll. Bull.* 62(2011) 1596–1597.

Ballent, A., Pando, S., Purser, A., Juliano, M.F., and Thomsen, L. (2013). Modelled transport of benthic marine microplastic pollution in the Nazare Canyon. *Biogeosciences*, 10(12): 7957–7970.

Browne, M.A., Galloway, T.S., and Thompson, R.C. (2010). Spatial patterns of plastic debris along estuarine shorelines. *Environmental Science & Technology*, 44: 3404–3409.

Colton, J.B., et al. (1974). Plastic particles in surface waters of the northwestern Atlantic. *Science*, 185: 491–497.

Day, R.H., Shaw, D.G., and Ingell, S.E. (1988). The quantitative distribution and characteristics of neuston plastic in the North Pacific Ocean. R.S. Shomura and M.L. Godfrey (ed.) Accessed 02/08/2021 @ https://swfsc.noaa.gov/publications/TM/SWFSC/NOAS-TM-NM FS-SWFSC-154_p247.PDF.

De, S.K. (1996). *Rubber Technologist's Handbook, Volume 1*. Akron, OH: Smithers Rapra Press.

Endo, S., Takizawa, R., Okuda, K., Takada, H., Chiba, K., Kanehiro, H., Ogi, H., Yamashita, R. and Date, T. (2021). Concentration of polychlorinated biphenyls (PCBs) in beached resin pellets: Variability among individual particles and regional differences. *Marine Pollution Bulletin*, in press.

Engler, R.E. (2012). The complex interaction between marine debris and toxic chemicals in the ocean. *Environmental Science and Technology*, 46(22): 12302–12315.

Eriksen, M., Maximento, N., Theil, M., Cummins, A., Latin, G., et al. (2013). Plastic marine pollution in the South Pacific Subtropical Gyre. *Marine Pollution Bulletin*, 68: 71–76.

Gabbatiss, J. (2018). Half of dead baby turtles found by Australian scientists have stomachs full of plastic. Accessed 02/08/2021 @ https://www.independent.co.uk/environment/turtles-plastic-pollution-deaths-australia-microplastic-waste-a8536041.html.

Graham, E.R., and Thompson I.T. (2009). Deposit- and suspension-feeding sea cucumbers (Echinodermata) ingest plastic fragments. Accessed 02/14/2021 @ https://www.cabdirect.org/cabdirect/abstract/search/20093325773.

Gregory, M.R. (2009). Environmental implications of plastic debris in marine settings-entanglement, ingestion, smothering, hangers-on, hitch-hiking and alien invasions. Accessed 02/08/2021 @ https://ncbi.nlm.nih.gov/pmc/articles/PMC2873013.

Hardy, J.T., Crecelius, E.A., Antrim, I.D., Kiesser, S.L., Broadhurst, V.L., Boehm, P.D., and Steinhauer, W.G. (1990). Aquatic surface contamination in Chesapeake Bay. *Marine Chemistry*, 28: 333–351.

ICS. (2011a). Polyethylene terephthalate. ICIS. Accessed 02/06/2021 @ http://ics.com/chemicals/polyethylene-terephthalate/.

ICS. (2011b). Polyethylene. ICIS. Accessed 02/06/2021 @ http://ics.com/chemicals/polyethylene/.

ICS. (2011c). Accessed 02/06/2021 @ http://ics.com/chemicals/polyvinyl-chloride/.

ICS. (2011d). Accessed 02/06/2021 @ http://ics.com/chemicals/polypropylene/.

ICS. (2011e). Accessed 02/06/2021 @ http://ics.com/chemicals/polystyrene/.

Maximenko, N. et al. (2012). Marine pollution in the South Pacific gyre. *Mar. Poll. Bull.* 68(2012) 71–76.

Moore, C. (2012). Plastic in Ocean. Accessed 02/12/21 @ https://wwwreserachgte.net/ublications/31802441.

Ng, K.L., and Obbard, J.P. (2006). Prevalence of microplastics in Singapore's coastal marine environment. *Marine Pollution Bulletin*, 52(7): 761–767.

NOAA. (2009). Facts about marine debris. Accessed 02/09/2021 @ http://marinedebris.noaa.gov.Marinedebris101/md101facts.html.

Parker, L. (2018). We depend on plastic. Now we're drowning in it. Accessed 02/08/2120. @ https://www.nationalgeographic.com/magazine/201806/plastic-planet-waste-pollution-trash-crisis.

Plastic Soup Foundation. (2019). What is plastic soup? Accessed 02/08/2021 @ https://www.plasticsoupfoundation.org/en/files/what-is-plastic-soup/.

Podsada, J. (2001). Lost sea cargo: Beach bounty or junk? *National Geographic News.* Accessed 03/09/09 @ http://news.National geographic.com/news/2001/06/0619_seacargo.html.

Pongracz, E. (2007). The environmental impacts of packaging. Accessed 02/13/2021 @ https://www.researchgate.net/publication/229796182.

Rios, I.M., and Moore, C. (2007). Persistent organic pollutants carried by synthetic polymers in the ocean environment. *Marine Pollution Bulletin*, 54(8): 1230–1237.

Spellman, F.R. (1996). *Stream ecology and self-purification*. Lancaster, PA: Technomic Publishing Company.

Spellman, F.R. (2014). *The Science of Water*, 3rd ed. Boca Raton, FL: CRC Press.

Teuten, E.L., Rowland, S.J., Galloway, T.S., and Thompson, R.C. (2007). Potential for plastics to transport hydrophobic contaminants. *Environmental Science & Technology*, 41: 7759–7764.

Thompson, R.C., Jon, A.W.G., McGonigle, D., and Russell, A.E. (2004). Lost at sea: Where is all the plastic? 304(5672): 838.

USEPA. (2017). Toxicological threats of plastic. Accessed 2/4/2021 @ https://www.epa.gov/trash-free-waters/toxicological-threats-plastic.

USEPA. (2020). Learn about ocean dumping. Accessed 2/4/2021 @ https://www.epa.gov/ocean-dumpng/learn-about-ocean-dumping#Before.

USEPA. (2021). Containers and packaging: Product-specific data. Accessed 02/06/2021 @ https://www.epa.gov/facts-and-figures-about-materials-waste-and-recycling/containers-and-packaging-product-specific-data.

6 E-Waste

Hey, Mate, There Is Gold and Silver and Gallium in Them Thar Devices!

—Frank R. Spellman

IT'S A FASHION STATEMENT

Well, it doesn't actually make one of those expected and rhythmic fashion catwalks. No narrow runways or ramps are used as the flat platform running into an auditorium or between a section of an outdoor seating area, and no models demonstrate clothing during a fashion show. None of that. Okay, then are we talking about scoring an exclusive for a particular designer only? Yes, could be. Can whatever that is being featured herein or therein be considered as haute couture? Depends on your point of view, with your view being the emphasis. Is this item typically featured in a look book—a collection of photos of the item, device, it, or whatever for fashion editors, buyers, clients, and special customers—for the viewer's perusal and to show the designer's model for the season, for the moment? How about sketches? Are there sketches of the product, the item, it, or whatever available for view, for study, for comparison?

The model, the item, it, or whatever being referred to here is high tech. The fact is tech has become another fashion item, and it can be said and easily noticed that many parts of the technology industry are now mutating into something with the anticipation and tempo of the fashion catwalk. And all this is quite confounding for traditional industries to find that what is hot today is colder than an iceberg and history tomorrow. The truth be known that the traditional industry personnel find the old rules of thumb are all of a sudden ancient and ignored.

So, what are we referring to here? What is it all about? Specifically, what is the item, device, it, or whatever? Well, be assured there are several items, devices, its, and whatevers that are being referred to here. However for clarity and to make the point transparent and as clear as a cloudless day, we will focus on one item, device, it, and whatever: the mobile phone.

In the 30 years that we have enjoyed wireless mobility with our cell phones, have you noticed or paid attention to the evolution of mobile, handheld, communication devices? Probably not. Well, each new generation of mobile technology has provided more bandwidth and more possibilities. Before fast-forwarding to the present status of mobile communication devices, it is important to list a brief review or timeline of the evolution of the mobile communication devices in use at the present time. The timeline is as follows:

1983—Motorola DynaTac 8000X
1989—MicroTac

DOI: 10.1201/9781003252665-6

1991—Orbitel TPU 900
1997—Seimens S10 (color made available)
Nokia 5100 series
1999—Nokia 7100
2000—Sharp J-SH04
2002—Sony T68i (w/camera)
2003—8100 Pearl
Sony 21020 (camera and video)
2007—Apple phone
2011–2014—Siei
Apple
2015–2018—iPhone 7 (size matters w/larger screen sizes)
Present—7 Pro device (all about speed)
5G—10 times faster than 4G

From the list above, it is obvious that mobile communication devices have continued to evolve or upgrade to the present time. Okay, technological progress is normal, natural, and to be expected. The drumbeat of innovation continues, sometimes at a startling pace. It is this startling pace of innovation that is the focus here. Simply, mobile communication devices are carry-everywhere, intimate devices. They express the user's personality, persona, and individuality. Newest, latest, slimmest, prettiest, shiniest, and fastest become urgent (the key word here is "urgent") point-of-purchase carrots or inducements.

Because of the rapid advance of mobile communication devices and their associated bells and whistles, it is incumbent upon the mobile device owner to obtain and possess the latest, greatest mobile phone.

Why?

Well, maybe the purchaser of the newest mobile phone wants to be a part of the so-called in-crowd, beau monde, the beautiful people, or the neighbor next door, but in this instance, whether it is the nerd who invents or possesses the newest model and/or style it does not matter. The latest, greatest mobile phone turns the possessors of such devices into copycats because possessing the latest great device is all about fashion.

At this point the reader may wonder what does all the preceding content have to do with waste? On the other hand, maybe some of the readers understand where the preceding discussion comes into play in regard to waste. Either way it is important to remember that we waste items when we no longer want them. We waste items when they are broken, no longer functional, or simply out of date. We also waste items when they are no longer fashionable. And that is exactly what occurs when our friends own the latest, greatest mobile phone and we do not. We simply must have the latest, greatest mobile phone—to fit in with the madding crowd and anyone else, so to speak.

The problem arises when the old mobile phone or device is no longer in fashion. What are we to do with the old one? Unfortunately, even though still functional and expensive when purchased, the dinosaur phone or electronic device must meet the same fate as the fossil. In the meantime, mobile phone users will continue to tether with their internet connection, smartphone zombie practices, and phubbing in favor of their newest state-of-the-art mobile phones.

THE 411 ON ELECTRONIC WASTE

According to USEPA (2020) report based on data from the Consumer Electronics Association, the average American household uses about 28 electronic products such as personal computers, mobile phones, televisions, and electron readers (e-readers). With an ever-increasing supply of new electronic gadgets, Americans generated about 2.7 tons of consumer electronics goods in 2018, representing about 1% of all municipal solid waste (MSW) generation (USEPA, 2020).

LIFE CYCLE OF ELECTRONICS

Raw Materials

Raw or virgin materials such as gold, silver, iron, palladium, copper, platinum, oil, and critical elements are found in a variety of high-tech electronics. They play crucial roles in products affecting our daily lives. These elements and materials are mined from the earth, transported, and processed. These activities use large amounts of energy and produce greenhouse gas emissions, pollution, and a drain on our natural resources. Source-reducing raw materials can save natural resources, conserve energy, and reduce pollution (Figure 6.1).

The problem with e-waste is that most people have no idea of what they are wasting. If you were to ask an average person if they have any idea of what their e-device is composed of, they probably will pause and scratch their heads and typically mutter something like "well, it is obvious that there is a lot of plastic involved and inside some electrical or electronic circuits and I am not sure what else is inside." It is the "I am not sure what else is inside" part of the comment that is concerning. If you were to explain that their e-device contained gold, usually $6 to $14 worth depending on the current spot price of gold, they might look at you and say, "Gee, I had no idea … but it is no matter, I have no idea how to recover the gold from the device … that is for someone else at the dump to figure out."

Sound familiar? Probably, that is if you ever ask someone or anyone who is dumping one e-device for another or whatever reason. So, what it comes down

FIGURE 6.1   Life cycle of electronics. Source: USEPA (2020).

to is probably based on two factors: factor one is ignorance. The average person has no idea of the value of the e-device he or she is trashing. Secondly, the person trashing the e-device could care less about the value of the e-device. "My neighbor has a new one … I want one of those too." This refrain is more common than we might imagine and is one of the main reasons why e-waste is a growing problem, worldwide.

The truth is few consumers of e-devices know the interior of their electronics have become thinly layered in gold; this is especially the case with computer desktops and laptops. This is why some computer makers are now offering to recycle our electronics.

You might wonder why so much of today's e-devices use gold instead of silver. Both are excellent conductors of electricity. Even though silver is less expensive than gold, the problem is corrosion, simply because silver corrodes in the environment (in the open air) while gold does not. Metallic corrosion is known and predictable, especially in silver. This is the principal reason gold is used in high-end e-devices. Along with being highly corrosion-resistant, gold is extremely malleable. Hence, while gold is not a better conductor of electrons than silver, it doesn't attract a sulfuric hue as silver does in the open air.

Why not use copper in e-devices, it is a lot cheaper than both gold and silver? The movement of electrons (electricity) in gold is many times faster than in silver; thus slices of gold often get selected for high-end electronics coatings.

The bottom line, e-waste is an increasing problem. USGS (2016) estimates that only about 13% of e-waste is being recycled. The electronic waste industry has become a globalized business. About 70% of e-waste is dumped into landfills, which it is typically sorted and sold for scrap metal. Sometimes shortsightedly, this e-waste is burned to extract valuable materials; this action is harmful to people and the surrounding environment and should not be practiced.

MINERALS IN MOBILE E-DEVICES

The plain truth is that high-technology devices such as mobile phones can't exist without mineral commodities. Mined and semi-processed materials (mineral commodities) make up more than 50% of all components in a mobile device—including its electronics, display, battery, speakers, and more (Ober et al., 2016). Figure 6.2 shows the ore minerals (sources) of some mineral commodities that are used to make the components of a mobile device. Table 6.1 lists examples of mineral commodities used in mobile devices.

DID YOU KNOW?

The glass video display compound of an electronic device, a cathode ray tube (CRT), is usually found in a computer or television monitor, and because it contains high enough concentrations of lead, the glass is regulated as hazardous waste when disposed of.

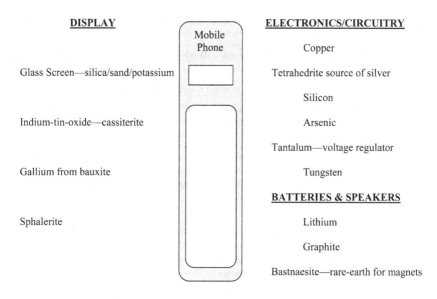

FIGURE 6.2   Minerals and derivatives in mobile phones. Source: USGS (2016).

E-Waste Capital of the World

In the Guangdong region of China, the massive electronic waste processing community of Guiyu is often referred to as the "e-waste capital of the world" (Basel Action Network, 2002; Roebuck, 2012; Slade, 2006). In the past, Guiyu was an agricultural community; however, in the mid-1990s it transformed into an e-waste recycling center involving over 75% of the local households and an additional 100,000 migrant workers (Wong, 2007). Laborers are employed in thousands of individual workshops to snip cables, pry chips from circuit boards, grind plastic computer cases into particles, and dip circuit boards in acid baths to dissolve the precious metals. Others work to strip insulation from all wiring in an attempt to salvage tiny amounts of copper wire (EDWD, 2012). Practices within these workshops such as uncontrolled burning, disassembly, and disposal have led to a number of environmental problems such as groundwater contamination, atmospheric pollution, and water pollution either by immediate discharge or from surface runoff (especially near coastal areas), as well as health problems, including occupational safety and health effects among those directly and indirectly involved, due to the methods of processing the waste.

Situated on the South China Sea coast, Guiyu consists of several villages. Six of these villages specialize in circuit-board disassembly, seven in plastics and metals reprocessing, and two in wire and cable disassembly. The environmental group Greenpeace sampled dust, soil, river sediment, and groundwater in Guiyu. They found very high levels of toxic heavy metals and organic contaminants (Seattle Times, 2012). The burning of plastics in the region has resulted in 80% of its children having dangerous levels of lead in the blood (Monbiot, 2009). One campaigner for the group found over ten poisonous metals such as lead, mercury, and cadmium.

TABLE 6.1

Examples of Mineral Commodities Used in Mobile Devices (USGS, 2016)

Mineral Commodity	Leading Global Sources	Mineral Source(s)	Properties of Commodity
Germanium	China	Sphalerite	Conducts electricity
Graphite	China, India	Graphite	Resists heat, conducts electricity, resists corrosion
Indium	China, Korea	Sphalerite	Transparent and conducts electricity
Lithium	Australia, Chile Argentina, China	Amblygonite, petalite, lepidolite, and spodumene	Chemically reactive and has a high performance-to-weight ratio
Platinum-group metals	South Africa, Russia, Canada	More than 100 minerals	Conducts electricity
Potassium	Canada, Russia, Belarus	Langbeinite, sylvite, and sylvinite	Strengthens glass
Rare-earth elements	China	Bastnaesite, ion adsorption clays, loparite, monazite, and xenotime	Highly magnetic; blue, green red, and yellow phosphors, and optical-quality glass
Industrial sand	China, United States	Silica sand	Gives glass clarity
Silicon	China	Quartz	Conducts electricity
Silver	Mexico, China, Peru	Argentite and tetrahedrite	Circuitry and conducts electricity
Tantalum	Rwanda, Brazil, Congo	Columbite and tantalite	Stores electrical charge
Tin	China, Indonesia, Burma, Peru	Cassiterite	Transparent and conducts electricity
Tungsten	China	Scheelite	Highly dense and durable for vibrator's weight component

Sadly, Guiyu is only one example of digital dumps, and similar places are currently found across the world such as in Nigeria, Ghana, and India (Greenpeace, 2021). Table 6.2 lists e-waste components and where they are found in electric/electronic appliances or devices and their adverse health effects.

RECYCLING SITE NEIGHBORS

Residents living around the e-waste recycling sites, even those who are not involved with the recycling, can face environmental exposure due to the food, water, and environmental contamination caused by e-waste because they can be easily exposed to or come in contact with the e-waste contaminated air, water, soil, dust, and food sources. In general, there are three main exposure pathways: inhalation, ingestion, and dermal contact (Grant et al., 2013).

TABLE 6.2
Hazardous Substances in E-Waste

E-Waste Component	Electric Appliances in Which They Are Found	Adverse Health Effects
Americium	Radioactive source in smoke alarms.	It is known to be carcinogenic (TOXNET, 2016).
Lead	Solder, CRT monitor glass, lead-acid batteries, and some formulations of PVC. A typical 15-inch cathode ray tube may contain 1.5 pounds of lead (Morgan, 2006).	Impaired cognitive function, behavioral disturbances, attention deficits, hyperactivity, conduct problems, and lower IQ (Chen, et al., 2011).
Mercury	Found in fluorescent tubes (numerous applications), tilt switches (mechanical doorbells, thermostats) (USEPA, 2009).	Health effects include sensory impairment, dermatitis, memory loss, and muscle weakness. Exposure in-utero causes fetal deficits in motor function, attention, and verbal domains (Chen et al., 2011).
Cadmium	Found in light-sensitive resistors, corrosion-resistant alloys for marine and aviation environments, and nickel-cadmium batteries.	The inhalation of cadmium can cause severe damage to the lungs and is known to cause kidney damage (Lenntech, 2014).
Hexavalent chromium	Used in metal coatings to protect from corrosion.	A known carcinogen after occupational inhalation exposure (Chen et al., 2011).
Sulfur	Found in lead-acid batteries.	Health effects include liver damage, kidney damage, heart damage, and eye and throat irritation. When released into the environment, it can create sulfuric acid through sulfur dioxide.
Brominated flame retardants	Used as flame retardants in plastics in most electronics.	Health effects include impaired development of the nervous system, thyroid problems, and liver problems (Birnbaum & Staskal, 2004).
Perfluorooctanoic acid	Used as an antistatic additive in industrial applications and found in non-stick cookware.	Studies on mice have found the following health effects: hepatoxicity, developmental toxicity, immunotoxicity, hormonal effects, and carcinogenic effects (Wu et al., 2012).
Beryllium oxide	Filler in some thermal interface materials such as thermal grease used on heat sinks for CPSs and power transistors (Becker et al., 2005), magnetrons, X-ray-transparent ceramic windows, heat transfer fins in vacuum tubes, and gas lasers.	Occupational exposures associated with lung cancer, and other common adverse health effects are beryllium sensitization, chronic beryllium disease, and acute beryllium disease (OSHA, 2016).
Polyvinyl chloride (PVC)	Commonly found in electronics and is typically used as insulation for electrical cables (Greenpeace, 2021).	In the manufacturing process, toxic and hazardous raw materials, including dioxins, are released. PVC such as chlorine tends to bioaccumulate (electronicstakeback, 2021).

Studies show that people living near e-waste recycling sites have a higher daily intake of heavy metals and a more serious body burden (toxic load). Potential health risks include mental health, impaired cognitive function, and general physical health damage (Song & Li, 2015). DNA damage was also found more widespread in all the e-waste exposed populations (i.e., adults, children, and neonates) than the populations in the control area (Song & Li, 2015). Experience has shown that DNA breaks can increase the likelihood of wrong replication and thus mutation, as well as lead to cancer if the damage is to a tumor suppressor gene (Liulin et al, 2011).

REFERENCES

Basel Action Network. (2002). Exporting harm: The high-tech trashing of Asia. Accessed 02/19/2021 @ http://ban.org/E-waste/technotrashfinalcomp.pdf.

Becker, G., Lee, C., and Lin, Z. (2005). Thermal conductivity in advanced CHIPS: Emerging generation of thermal greases offers advantages. Accessed 02/11/2021. @ https://web.archive.org/web/20000621233638/http:www.apmag.com/.

Birnbaum, L.S., and Staskal, D.F. (2004). Brominated flame retardants: Cause for concern? *Environmental Health Perspectives*, 112(1): 9–17.

Chen, A., Dietrich, K.N., and Huo, X. (2011). Developmental neurotoxicants in E-waste: An emerging health concern. Accessed 02/11/2021 @ https://ncbi.nim.nih.gov/pmc/acticl es/PMC3080922.

Electronicstakeback. (2021). Flame retardants & PVCs in electronics. Accessed 02/11/2021 @ http://www.electronicstakeback.com/toxics-in-electroncis/flame-retardants-pvc-and-electroncs.

EWDW. (2012). Electronic waste dump of the world. Accessed 02/19/2021 @ http://sometime s-interesting.com/2011/07/17/electronic-waste-dump-of-the-world/.

Grant, K., Goldizen, F.C., Sly, P.D., Brune, M.-N., Neira, M., van den Berg, M., and Norma, R.E. (2013). Health consequence of exposure to e-waste: A systematic view. *The Lancet Global Health*, 1(6): 350–361.

Greenpeace. (2021). Why BFRs and PVC should be phased out of electron devices. Accessed 02/11/2021 @ http://www.greenpeace.org/archive-international/en/campaigns/detox/ electronics/the-e-waste-problem/what-s-in-electronc-devices/bfr-pvc-toxic/.

Lenntech. (2014). Cadmium (Cd)-chemical properties, health and environmental effects Accessed 01/22/2021 @ http://web.archive.org/web/20140515101400/http://www.l enntech.com/periodic/elements/cd.htm#ixzz1MpuZHWfr.

Liulin, L. et al. (2011). Significance of tumor suppression genes. Accessed 1/17/21 @ https:// pubmed.ncbi.nim.nih.gov/27179964.

Monbiot, G. (2009). From toxic waste to toxic assets, the same people always get dumped on. *The Guardian*. London. Accessed 02/18/2021 @ https://www.theguardian.com/co mmentisfee/cif-green/2009/sep/21/global-fly-tipping-toxic-waste.

Morgan, R. (2006). Tips and tricks for recycling old computers. Accessed 02/19/2021 @ http:// www.smartbiz.com/acrticle/articleprint/1525-1/58.

Ober, J. et al. (2016). A world of minerals in your mobile device. Accessed 02/12/21 @ https:// pubs.usgs.gov/gip/0167.

OSHA. (2016). Health effects. Accessed 02/02/2021 @ https://www.osha.gov/SLTC/bery llium/healtheffects.html.

Roebuck, K. (2012). Electronic waste: High-impact strategies. Accessed 02/19/2021 @ https:// books.google.com/books?id=RjYPBwAAQBAJ&q=Activists+push+for+safer+e-recy clin&pg=PA8.

Seattle Times. (2012). E-waste dump of the world. Accessed 02/19/2021 @ http://seatteltimes. com/htiml /nationworld/2002920133_ewaste09.html.

Slade, G. (2006). Computer age leftovers. Accessed 02/19/2021 @ http://www.denverpost. com/perspective/ci_3633138.

Song, Q. and Li, J. (2015). A review of human health consequences of metal exposure to e-waste in China. *Environmental Pollution*, 196: 450–461.

TOXNET. (2016). Americium, radioactive. Accessed 02018/2021 @ http://toxnet.nim.nih. gov/cgi-bin/sis/search/a?dbs+hsdb:@DOCNO+7383.

USEPA. (2009). Question 8. Accessed 02/05/2021 @ http://www.epa.gov/dfe/pubs/comp-dic-ica-sum/ques8.pdf.

USEPA (2020). Cleanup of electronic waste. Accessed 02/11/21 @ https://www.epa.gov/int ernationalcooperation.

USGS. (2016). *Obsolete Computers, "Gold Mine," or High-Tech Trash?* Washington, DC: U.S. Department of Interior—U.S. Geological Survey.

Wong, M.H. (2007). Export of toxic chemicals-A review of the case of uncontrolled electronic-waste recycling. Accessed 02/19/2021 @ https://repository.hkbu.edu.hk/cgi/ viewcontent.cgi?article=1000&context=cies_ia.

Wu, K., Xu, X., Peng, L., Liu, J., Guo, Y., and Huo, X. (2012). Association between maternal exposure to perfluorooctanoic acid (PFOA) from electron waste recycling and neonatal health outcomes. *Environmental International*, 41: 1–8.

7 Food Waste

In the United States, over one-third of all available food goes uneaten through loss or waste. When food is tossed aside, so too are opportunities for improved security, economic growth, and environmental prosperity. United States Departmen of Agriculture (USDA) is uniquely positioned to help address the problem of food loss and waste through its programs, policies, and guidance.

—USDA, 2014

THE ONE-THIRD (1/3) FACTOR

The world produces enough food to feed a population total of almost double the number of humans on Earth today. But at the same time, we have large numbers within the world's population number that are malnourished or starving to death. Some blame the malnutrition and starvation problem on over population; these soothsayers champion the words of Malthus: "The power of population is so superior to the power of the Earth to produce subsistence for man that premature death must in some shape or other visit the human race." The truth be told is that we could solve the malnutrition and starvation problems currently prevalent in certain global regions by not wasting the food we have. One-third and maybe more of the food we produce each year is never eaten—in this book, it is known as the one-third factor.

Before moving on in this presentation it is important right up front, so to speak, to distinguish between two often-confused and commonly interchanged terms that for many have blurred meanings. The two terms are *food loss* and *food waste* and they are not the same—there is a difference in meaning between them. The difference is based on where the problem occurs along the food's journey from farm to consumer (see Figure 7.1). Food loss basically occurs on the farm, during storage, and during transport; it can be said that it is food lost from farm to fork. While food waste occurs in restaurants, in households, and in stores—from shelve to table. Both food loss and food waste can simply be defined as food that is not eaten.

Note: Whether we call food as a "loss" or a "waste" for the purpose of discussion herein is not important. Therefore, the information that follows uses the terms interchangeably to deliver the crux of the presentation and food waste/loss importance as a major ingredient or subset of waste in general.

FOOD NOT EATEN

At this particular moment in this discussion the logical question might be: When does food lost become food waste? Well, simply, food loss represents the edible

FOOD LOSS	FOOD WASTE
On the farm	In restaurants
During storage	In households
During transport	In stores

FIGURE 7.1   Food loss vs food waste.

amount of food, post-harvest, available for human consumption but not consumed for any reason. Food loss includes:

- Loss from molds, pests, or inadequate climate control
- Cooling loss and natural shrinkage (i.e., moisture loss)
- Food waste (a good example is food left on a plate and dumped into a trash can or garbage disposal)

So most of us in the United States and other countries have experienced food loss that becomes food waste. The question is why does food loss occur? Food loss happens for many reasons. Usually, food loss is accompanied or perpetrated simply because of loss such as spoilage which occurs at every stage of the production and supply chain. Between the farm field and retail shelves, food loss can arise from problems during drying, million transporting, or processing that expose food to damage by insects, rodents, birds, olds, and bacterial. At the grocery store or other food supply outlet, equipment malfunction (the cold storage malfunctions or fails), over-ordering, and culling (discarding, rejection, removing, or getting rid of) of blemished produce can result in food loss. Consumers also add to food loss when they cook more than they need and toss out the extras. Keep in mind that there are some good economic reasons for some food loss, such as to ensure food safety and to avoid costly medical and legal expenses.

The question shifts to exactly how much food loss is there? The USDA's Economic Research Service (ERS, 2020) approximates that a total of 133 billion pounds, or 31%, of the 430 billion pounds of accessible food supply at the retail and consumer levels went uneaten, with an estimated retail value of $162 billion. This translates into 141 trillion calories (Kcal) of food available in U.S. food supply but not consumed in 2010. On a per-capita basis, this totaled approximately 1.2 pounds of food per person per day, with a retail value of over $1.40.

ERS RESEARCH FINDINGS

Fresh Vegetable Losses

While conducting research on various aspects of food loss problems in the United States, the ERS studied the amounts and rates of retail food loss and discovered that food loss varies by type of fresh vegetables.

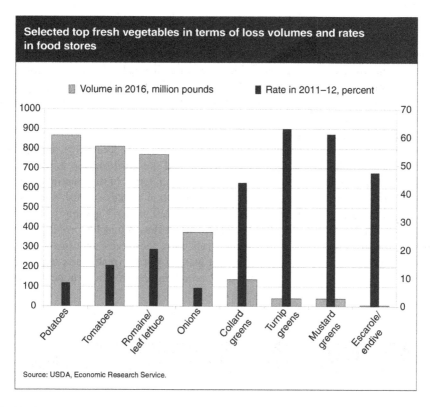

FIGURE 7.2   Top fresh vegetables in terms of loss volumes and rates in food stores. Source: USDA (2020).

In general, retail-level food loss occurs when grocery retailers remove misshaped produce items, dented cans, overstocked holiday foods, and spoiled foods from their shelves. Estimate of the average supermarket loss rate was 11.6% for 31 fresh vegetables, with the highest loss rate for turnip greens, followed by mustard greens, and escarole/endive (see Figure 7.2). To determine a reasonable estimate of food store loss rates for fresh produce, using representative sampling techniques, ERS developed a comparison scheme whereby data on pounds of shipments received with pounds purchased by consumers from 2,900 supermarkets in 2011–2012 in the United States were compared. After determining loss rates, ERS researchers applied these loss rates to 2016 quantities of fresh vegetables available for sale in retail stores to estimate retail-level food loss.

Note that while the loss rates for potatoes, tomatoes, and romaine and leaf lettuce are lower than turnip and mustard greens, their sales volumes are higher accounting for 35% of food store fresh vegetable sales. In 2016, supermarket loss for the 31 fresh vegetables totaled 6.2 billion pounds per year, or 5 billion pounds per year after removing the weight of nonedible peels, stalks, and other parts. Keep in mind that losses for fresh produce and other foods also occur in homes and eating places when food spoils or is served but not eaten (plate waste) (USDA, 2020).

Fresh Fruit Loss

Rates of supermarket loss for 24 fresh fruits, based on a representative sample of 2,900 U.S. supermarkets in 2011–2012, were estimated by comparing pounds of shipments received by grocers with pounds purchased by consumers. Low- to high-loss rates ranged from 4.1% for bananas to 43.1% for papayas (see Figure 7.3). Overstocking as well as greater perishability may contribute to higher loss rates (USDA, 2020).

ERS researchers applied the 2011–2012 loss rates to estimate retail-level loss rates of fresh fruits available for sale in retail stores. They found that pineapples had the second-highest loss rate, and apricots had the third-highest loss rate for fresh fruits. Pineapples also ranked relatively high in terms of the amount of retail loss in 2016 (719 million pounds) due to the 2.2 billion pounds of fresh pineapples available for sale in retail stores that year. The researchers' assumptions about fresh watermelon and apples were validated when they found loss volumes were highest for fresh watermelon and apples, reflecting the large quantities available for sale by retailers. In 2016, supermarket loss for the 24 fresh fruits totaled 6.7 billion pounds, or 4.7 billion pounds after removing the weight of nonedible pits and peels (USDA, 2020).

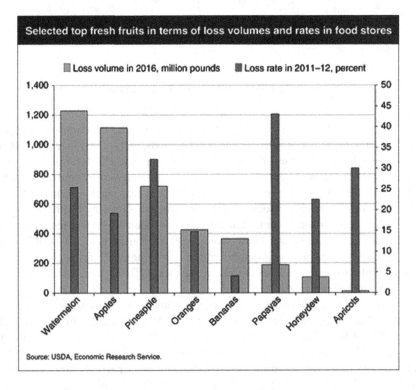

FIGURE 7.3   Fresh fruit loss 2011–2012. Source: USDA (2020).

THE PSYCHOLOGY OF FOOD WASTE

Along with the rejection of suboptimal food described above, consumers indirectly waste food for psychological reasons.

Psychological reasons?

Yes. Absolutely. Simply, and according to Aschemann-Witzel et al. (2019) "food waste at the consumer level is caused by the consumer-related factors of motivation, tradeoff, and capability, as well as by the context-related factors of social influence, purchase context, and the macroenvironment." Basically, the point being made here is that a large portion of food waste is avoidable, and it is often due to suboptimal food (as described earlier). The reality is it is a matter of consumer perception—they feel that suboptimal food is either of lesser value or unsafe to eat. People usually do not purchase items that are damaged. This is the same feeling many purchasers have when they view food packaging that has been damaged, opened, or discolored. Note that the purchaser makes decisions related to the condition of food and food packaging both at the store and at home. When a person opens his or her refrigerator and removes a bottle of milk and notices that the printed use before date is long past, they are likely to pour the container contents down the drain. This is a classic instance of food waste and is even more common than you might think. Milk is not the only food item that goes down the drain or into the garbage can.

Another factor related to psychology-based food waste is value and cost. Consumers shy away from purchasing anything that is either too expensive or is not worth the price being charged. When choosing food to purchase, it is common practice for the consumer to choose food stuffs that are reasonably priced and in good condition. Supermarket and suppliers are well aware of the psychology of food. They know that if they attempt to sell a damaged or an out-of-date food stuff, the consumer will reject those items; thus, those food items judged unsalable are summarily trashed—wasted.

The bottom line: Consumer perception and retailer reactions are major contributors to wasting food.

DID YOU KNOW?

The process of dehydrating foods into powders raises the question: In the process, what is wasted, if anything? Partial or complete drying of food stuffs is performed to inhibit microbial growth, to preserve quality, to allow their applications in various foods, and recently to convert to pill or capsule form as a simpler way to get veggies and fruits; these capsules typically contain several veggies and fruits, grains, greens, antioxidants, and digestive enzymes. These dehydrated food stuffs are daily gaining in popularity. To answer the question as to what is wasted in the dehydration process of veggies and fruits, the main waste product is water. However, if the veggie or fruit is low in moisture content, dehydration can lead to chemical degradation and nonenzymatic browning.

—Spellman and Price-Bayer, 2019

VEGETABLE LOSSES AFTER HARVESTING

After harvesting vegetables and fruits, post-harvest losses of vegetables and fruits occur at all points in the food value chain from production in the field to the food on a fork for consumption. The *food value chain* is a relatively new model of organization that is beginning to pop up in the agribusiness sector that seeks to merge social mission objectives with core business operating principles—these business arrangements are distinguished by their commitment to transparency, collaborative business planning and exchange of market intelligence and business know-how among chain partners, and their interest in developing business strategies and solutions that yield tangible benefits to each participant in the system (USDA, 2014). Note that external factors that have contributed to the rise of food value chain enterprises recently include the growing segmentation of the consumer market, escalating demand for specialized, highly differentiated food products—even at higher prices points—and the increasing appeal for food items that are produced in accordance with desired social or environmental welfare standards (USDA, 2014). With regard to post-harvest activities of vegetables, after the harvest, handling, storage, processing, packaging, transportation, and marketing are involved (Mrema and Rolle, 2002).

Experience has shown that loss of vegetables and fruits is a major problem in the post-harvest chain. Causal factors range from growing conditions to handling at the retail level. It is important to point out that not only are losses clearly a waste of food, but they also represent a similar waste of human effort, farm inputs, livelihoods, investments, and scarce resources such as water (World Resources Institute, 1998), fertilizer, and farm equipment wear and tear including wasted farm machinery fuel. One of the headache-makers for researchers trying to measure post-harvest losses of vegetables and fruits is that measurement and magnitude of the losses in any form is difficult. There are instances when everything harvested by a farmer ends up being sold to consumers. However, in others, losses and/or wastes may be considerable. Sometimes, losses may be total (100%); for example, when there is a price collapse, it would cost the farmer more to harvest and market the produce than to turn it back into the ground. Because of this and other issues use of average loss figures is often misleading. Not only can there be losses in quantity but also losses in quality—based on measures of the price obtained and the nutritional value.

Note that on-farm causes of loss are not limited to one or two factors but also include the type of soil the crop was grown on and the handling of produce when it reaches the shopper. Moreover, pre-harvest production practices may seriously affect post-harvest returns. Good crop husbandry is important for shrinking losses. Good husbandry, for example, entails ensuring plants have a continuous supply of water for transpiration and photosynthesis. Irrigation must be closely monitored because too much water, irregular water supply, and too little water can lead to decay and other problems. Proper application of plant food is also important because too little plant food leads to lowering the quality of fresh produce, bring about stunted growth and/ or discoloration of leaves, and abnormal ripening—too much plant food (fertilizer) can harm the development and post-harvest condition of produce. Another issue that can't be ignored is weeds. Weeds compete with crops for nutrients and soil moisture. Another major loss factor is in-field decaying plant residues (FAO, 1989; Kader, 2005; Lopez-Camelo, 2004).

CAUSES OF AFTER-HARVEST LOSSES

Because living plant parts such as fruits and vegetables contain 65–95% water, they die and decay when water is not available or provided; this is also the case when plant food is not available. Other injuries to fruits and vegetables occur when temperatures are too high or when humidity is too low. Any practice or natural occurrence (e.g., storms) that causes physical damage is also an issue (Onwude et al., 2020). These injuries result from natural occurrences and reckless handling, causing internal bruising, cracking, and skin breaks, as a result of increasing water loss.

Note that after harvest, fruits and vegetables continue to live via respiration which is an ongoing process until all supplies are exhausted. This is an important factor because in order to keep fruits and vegetables fresh, it is essential that respiration is kept going on for as long as possible. Respiration in fruits and vegetables is important because it sets free energy which can be used for their internal processes. As long as the respiration process continues, the fruits and vegetables can be kept sound. When the fruits and vegetables are harvested, they are basically on their own, meaning that being no longer root-connected, they no longer receive nutrients; thus, when the internal supply is gone, the fruits and vegetables deteriorate.

DID YOU KNOW?

Respiration in fruits and vegetables is a very common chemical reaction. However, respiration in fruits and vegetables is not unique, it takes place in all plants and animals. The chemical reaction can be represented as follows:

$$C_6H_{12}O6 + 6O_2 \rightarrow 6CO_2 + 6H_2O + ENERGY$$

The bottom line: The respiration rate of a product determines how fast the chemical process occurs.

Produce must be kept in a moist environment. This is the case because fresh produce continues to lose water after harvest. Water loss causes loss of weight and shrinkage. The rate at which water is lost varies according to the product. Those vegetables that are leafy lose water quickly because they have thin skin with many pores. On the other hand, potatoes have thick skin with few pores. However, whatever the product, to extend shelf or storage life, the rate of water loss must be minimal.

Another after-harvest loss of fruits and vegetables is caused by diseases. The losses are brought about by fungal and bacterial diseases and not by viral diseases. When deep decay occurs, infected produce is unsalable. This is often the result of in-field infection of the produce before harvest. However, purchasers typically purchase fruits and vegetables based on the products' appearance. When a disease affects the product's surface its quality is diminished, making it likely that purchasers move on to other products. Also, when fruits and vegetables are blemished, they are unacceptable, and lowering the sale price is an option but is rarely a reason for purchasers to buy; quality in appearance is the bottom line.

Probably the most common problem affecting the quality of after-harvest fruits and vegetables is their tendency to ripen only while still attached to their stems. Eating quality suffers if they are harvested before fully ripe as their acid and sugar content does not increase further. It is true that some fruits can be harvested before ripening (e.g., bananas, melons, papayas, and tomatoes) so long as they are mature. In commercial fruit marketing the rate of ripening is controlled artificially, this enables the transport and distribution to be carefully planned. In preparing some fruits for sale, grocers have learned to use ethylene gas to aid in the ripening process. Ethylene gas is produced in most plant tissues and is important in starting the ripening process. On the other hand, ethylene gas damages leafy vegetables if stored with ripening fruit (Kader, 2005; Lopez-Camelo, 2004).

Fruits and vegetables are very vulnerable to mechanical damage, and it can occur at any time during the marketing chain. Poor harvesting practices such as the use of filthy cutting knives, unsuitable containers used at harvest time or during the marketing process, overpacking or underpacking of containers, and the careless handling of containers can all cause mechanical damage. Resultant damage can include splitting of fruits, internal bruising, superficial grazing, and crushing of soft produce. Thus, poor handling can result in the development of entry points for molds and bacteria, increased water loss, and increased respiration rate (Dixie, 2005).

DID YOU KNOW?

Unsuitable containers used at harvest time or during the marketing process are those that can be easily squashed or have splintered wood, sharp edges, or poor nailing or fastening.

Temperature extremes can also contribute to fruit and vegetable damage. Levels of tolerance to low temperatures are important when cool storage is foreseen. Between 0°C and −2°C, all produce will freeze. Improper temperature control can lead to significant losses.

Oftentimes, fruits and vegetables are exposed to filthy contaminants. Many of these impurities are introduced after harvest by the use of contaminated field boxes; dirty wash water; decaying, rejected produce lying around packing houses; and unhealthy produce contaminating healthy produce in the same packages.

Fruit and vegetable losses during transportation can be high. Damage usually occurs during the loading and unloading of packed produce and where and when the procedures are conducted in a careless manner. Also, during transport the produce might be exposed to vibration and shaking generated by the vehicle, especially on bad farm roads. Poor storage of the produce is another issue, which is especially the case whenever the packages are squeezed into the vehicle in order to maximize space usage and increase revenue for the transporters. Closed vehicles with no to little ventilation and/or overexposure to the sun while hauling produce are also a cause of food loss and waste. The overheating of the produce leads to decay and increases the rate of water loss. Breakdown of transportation vehicles can be a significant causal factor related to food loss and waste—the produce can spoil while waiting for repairs.

Food loss and waste is not limited to the farm and/or to transportation modes. At the retail marketing stage, losses can be (and are) significant. Poor-quality markets often provide little protection for the produce against the elements, leading to rapid produce deterioration and subsequent throw-away. Experience has demonstrated that when produce is sorted to separate the salable from the unsalable, the result can be high percentages being discarded, and there can be high weight loss from the trimming of leafy vegetables. The old produce is often discarded anytime fresh supplies arrive; thus, this is another factor or practice directly related to food loss and waste.

As discussed to this point it is clear that there are many means and events and practices that can lead to fruit and vegetable loss and waste. But it is important to point out that these foodstuffs are not the only types of food products that are lost or wasted for whatever reason or occurrence. Consider, for example, grains and grain products. These grains and grain products are exposed to many of the same losses and wasting as are the fruits and vegetables. Grains may be lost and wasted during pre-harvest, harvest, and post-harvest stages. Pre-harvest losses occur before the process of harvesting begins, and may be due to insects, weeds, and rusts.

The potential for an actual grain loss and subsequent waste production is a real, ongoing, and constant concern for farmers and others. Let's look at wheat diseases and pests, for example. The following provides a representative sample of wheat diseases, pests, physiologic and genetic disorders, and mineral and environmental stresses that are currently of major concern to wheat farmers (USDA, 2021b).

FUNGAL DISEASES

- Leaf Rust (Brown Rust)
- Stem Rust (Black Rust)
- Stripe Rust (Yellow Rust)
- Common and Dwarf Bunt (Stinking Smut—rotten fish odor)
- Karnal Bunt (Partial Bunt)
- Loose Smut
- Flag Smut
- Powdery Mildew
- Tan Spot (Yellow Leaf Spot or Blotch)
- Eyespot (Strawbreaker)

BACTERIAL DISEASES

- Bacterial Black Chaff and Bacterial Stripe
- Basal Blume Rot and Bacterial Leaf Blight
- Bacterial Spike Blight (Yellow Ear Rot)

VIRAL DISEASE

- Barley Yellow Dwarf

INSECT PESTS

- Aphids
- Thirds
- Wheat Stem Maggot
- White Grubs
- Mites

NEMATODES (WORMS)

- Seed Gall Nematodes (Wheat Nematode or Ear Cockle)
- Cereal Cyst Nematode
- Root Knot Nematode

PHYSIOLOGIC AND GENETIC DISORDERS

- Physiological Leaf Spot
- Melanism and Brown Necrosis (False Back Chaff)
- Genetic Flecking (Blotching)

MINERAL AND ENVIRONMENTAL STRESSES

- Nitrogen, Phosphorus, and Potassium Deficiencies
- Minor Element Deficiencies
- Aluminum Toxicity
- Salt Stress
- Moisture Stress
- Heat Stress
- Herbicide Damage
- Frost Damage

During harvest, grain losses occur from beginning to end and are primarily caused by losses due to shattering (i.e., the undesirable scattering of seeds upon maturation (ripening)). The post-harvest losses include grain threshing, winnowing, and drying. Also, losses occur along the chain during transportation, storage, and processing.

DID YOU KNOW?

During the shelling process, the stripping of maize grain from the cob, losses can occur when mechanical shelling is not followed up by hand-stripping of the grains that are missed. During shelling, certain workers can damage the grain, making insect penetration easier.

REFERENCES

Aschemann-Witzel, J., de Hooge, I.E., and Almli, V.L. (2019). Saving food: Production, supply chain, food waste and food consumption. *Science Direct*, 22: 347–368.

Dixie, G. (2005). Horticultural marketing. Accessed 02/28/2021 @ ftp://ftp.fao.org/docrep/fao/008/a0185e/a0185e00.pdf.

Farm and Agriculture Org (FAO-UN). (1989). FAO prevention of post-harvest food losses: Fruits, vegetables and root crops – A training manual. Accessed 02/28/2021 @ http://www.fao.org/docrep/T0073E/T0073E00.htm#Contents.

Kader, A.A. (2005). Increasing food availability by reducing postharvest losses of fresh produce. Accessed 02/28/2021 @ https://web.archive.org/web/20100613050639/http://postharvest.ucdavis.edu/datastorefiles/234-528.pdf.

Lopez-Camelo, A. (2004). Manual for the preparation and sale of fruits and vegetables—from farm to market. Accessed 02/28/2021 @ ftp://ftp.fao.org/docrep/fao/008/y4893e/y4893e00.pdf.

Mrema, C.G. and Rolle, S.R. (2002). Status of the postharvest sector and its contribution to agricultural development and economic growth, *9th JIRCAS International Symposium—Value Addition to Agricultural Product*, pp. 13–20. Accessed 02/28/2021 @ http://www.fao.org/ag/ags/subjects/en/harvest/docs/Mrema_Rolle.pdf.

Onwude, D.I., Chen, G., Eke-emezie, N., Kabutey, A., Khaled, A., and Strum, B. (2020). Recent advances in reducing food losses in the supply chain of fresh agricultural produce. Accessed 02/28/2021 @ https://doi.org/10.3390%2Fpr8111431.

Spellman, F.R. and Price-Bayer, J. (2019). *Regulating food additives: The good, bad and the ugly*. Latham, MD: Bernan Press.

USDA. (2014). Cass County FSA Updates. Accessed 02/04/2021 @ https://www.fsa.usda.gov/internet/fsa.

USDA. (2020). Food loss. Accessed 02/22/2020 @ https://www.ers.usda.gov/data-products/food-availablity-per-capita-data-system/food-loss/.

USDA. (2021a). The impact of food waste. Accessed 02/20/2021 @ https://www.usda.gov/foodloss and waste.

USDA. (2021b). Wheat diseases and pests: a guide for field identification. Accessed 02/27/2021 @ https://wheat.pw.usda.gov/ggpages/wheatpests.html.

World Resources Institute. (1998). Disappearing food: How big are postharvest losses? Accessed 02/28/2021 @ https://web.archive.org/web/20100508154316/http://earthtends.wri.org/pdf_libraby/feature/agr_fea_disappear.pdf.

8 Solid Waste, Landfills, and Leachate

SOLID WASTE

Humans have always been known for their garbage. Ancient garbage dumps (called *middens*) provide us with a wealth of information on our human ancestors and their lives. From the historical perspective, this human garbage trail has allowed archaeologists to study humans from their earliest days, discovering many fascinating facts about them (us). From the garbage record, for example, we've determined that at every stage, humans have always lived with enormous amounts of garbage, underfoot and all around them. Still true in the poorest parts of the developing countries, this all-too-human habit poses a significant hazard to human health and to the environment, today and always.

Ah! But don't panic. Modern Man can solve any problem. Right?

Is this really the case? Let's look at the record—the record we will leave behind us—maybe it will provide us with some insight or answers.

By the 20th century, most industrialized countries (with their modern approaches to sanitation) removed garbage and other human waste from many living and working environments. But how about the other environment?—the one we abuse or rarely think about (the one that sustains our lives)—in which we live with its air, soil, and water?—all critical to our very existence?—the natural world?

We stated earlier that people can put up with just about anything, until it displeases them. Then, of course, the source of displeasure must be removed—the old "out of sight out of mind" syndrome. So, what did we do? What do we do today?

We transferred or diverted our garbage from our immediate living and working area into waterways, we heaped and often burned it in garbage dumps (today called *surface impoundments*), or we dumped it into areas called *landfills* (often, former *wetlands* were filled in for anticipated future use).

The result has been so massive as to overwhelm and sometimes directly kill life in local environments. Each year, in the United States alone, about 10 billion metric tons of non-agricultural solid waste is generated. Municipal solid waste alone accounts for more than 150 metric tons each year. An average US citizen discards about four pounds of waste each day—that's nearly 1,500 pounds per person, each year.

But what is new and threatening about modern garbage is not entirely the amount (great as it is), but its toxicity and persistence. Most waste in earlier times was *biodegradable*—that is, it could and did break down in the environment as part of natural processes (via biogeochemical cycles). However, today, humans routinely use products made from or that produce toxic chemicals, and many other hazardous

DOI: 10.1201/9781003252665-8

substances. Many of these are poisonous to start with; others become poisonous under certain situations—for example, when they are burned, or when they come into contact with certain other chemicals to form a chemical brew of unknown toxicity. This chemical brew also enters the food chain and is passed along in concentration in the bodies of larger organisms. Many of these waste products, pesticides and plastics especially, persist in the environment for years, decades, and beyond.

Nearly 20 years ago, the *Wall Street Journal*, the *Washington Post*, and the *Christian Science Monitor* all called the garbage disposal situation in America a "crisis" (Peterson, 1987; Tongue, 1987; Richards, 1988). Have we weathered the crisis? Can it be solved, or will we continue to expand the historical garbage trail, leaving future archaeologists a record to study that will paint our generation as one of folly, of total disregard, of deliberate misuse—capable only of poor judgment and for whom conscience is for the other guy, for other generations?

—**Spellman and Whiting (2006)**

OUT OF SIGHT, OUT OF MIND

In this section, we discuss a growing and significant problem facing all of us: *anthropogenically produced solid wastes*. Specifically, we ask and wonder what are we going to do with all the solids wastes we generate? What are the alternatives? What are the technologies available to us at present to mitigate the waste problem, a problem that grows with each passing day?

Before beginning our discussion, we focus on an important question: When we throw waste away, is it really gone? Remember, though we are faced today and in the immediate future with growing mountains of wastes we produce (and we are running out of places on earth to dispose of them), an even more pressing two-fold problem is approaching—one related to the waste's toxicity and persistence.

It is important to think about the persistence of the wastes that we dispose of. For example, when we excavate a deep trench and place within it several 55-gallon drums of liquid waste, then bury the entire sordid mess, are we really disposing of the waste in an Earth-friendly way? Are we disposing of it permanently at all? What happens a few years later when the 55-gallon drums corrode and leak? Where does the waste go? What are the consequences of such practices? Are they insignificant to us today because don't worry they are tomorrow's problems? We need to ask ourselves these questions and determine the answers now. If we are uncomfortable with the answers we come up with now, shouldn't we feel the same about the answers someone else (our grandchildren) will have to come up with later?

Waste is not easily disposed of. Oh sure, we can hide or mask it—out of sight out of mind. Remember, we soon forget things no longer present or visible. With regard to persistent solid waste, we can move it from place to place. We can take it to the remotest corners of the earth. But because of its persistence, waste is not always gone when we think it is. It has a way of coming back, a way of reminding us—a way of persisting.

In this section, we define and discuss solid wastes, landfills, and leachate. In particular, we focus a significant portion of the discussion on solid wastes, *Municipal*

Solid Wastes (MSW), because people living in urban areas where many of the problems associated with solid waste occur are the generators of these wastes.

SOLID WASTE REGULATORY HISTORY IN THE UNITED STATES

For most of the nation's history, municipal ordinances (rather than federal regulatory control) were the only solid waste regulations in effect. These local urban governments controlled solid waste almost from the beginning of each settlement—because of the inherent severe health consequences derived from street disposal. Along with prohibiting dumping of waste in the streets, municipal regulations usually stipulated requirements for proper disposal in designated waste dump sites, and mandated owners to remove their waste piles from public property.

The federal government did not begin regulating solid waste dumping until the nation's harbors and rivers were either overwhelmed with raw wastes or headed in that direction. The federal government used its constitutional powers under the Interstate Commerce Clause of the constitution to enact the Rivers and Harbors Act in 1899. The US Army Corps of Engineers was empowered to regulate and, in some cases, prohibit private and municipal dumping practices.

Not until 1965 did Congress finally get into the picture (as a result of strong public opinion) by adopting the Solid Waste Disposal Act of 1965, which became the responsibility of the US Public Health Service to enforce. The intent of this act was to (Tchobanoglous et al., 1993):

1. Promote the demonstration, construction, and application of solid waste management and resource recovery systems that preserve and enhance the quality of air, water, and land resources.
2. Provide technical and financial assistance to state and local governments and interstate agencies in the planning and development of resource recovery and solid waste disposal programs.
3. Promote a national research and development program for improved management techniques; more effective organizational arrangements; new and improved methods of collection, separation, recovery, and recycling of solid wastes; and the environmentally safe disposal of nonrecoverable residues.
4. Provide the promulgation of guidelines for solid waste collection, transport, separation, recovery, and disposal systems.
5. Provide training grants in occupations involving the design, operation, and maintenance of solid waste disposal systems.

After Earth Day 1970, Congress became more sensitive to waste issues. In 1976, Congress passed solid waste controls as part of the *Resource Conservation and Recovery Act* (RCRA). "Solid waste" was defined as any garbage, refuse, sludge from a waste treatment plant, water supply treatment plant, or air-pollution control facility, and other discarded material.

In 1980, Public Law 96-510, 42 USC Article 9601, the Comprehensive Environmental Response, Compensation and Liability Act (CERCLA) was enacted to provide a means of directly responding, and funding the activities of response, to problems at uncontrolled hazardous waste disposal sites. Uncontrolled MSW landfills are facilities that have not operated or are not operating under RCRA (USEPA, 1989).

Many other laws that apply to the control of solid waste management problems are now in effect. Federal legislation and associated regulations have encouraged solid waste management programs to be implemented at the state level of government. Apparently, legislation will continue to be an important part of future solid waste management.

SOLID WASTE CHARACTERISTICS

Solid waste (also called *refuse, litter, rubbish, waste, trash,* and (incorrectly) *garbage*) refers to any of a variety of materials that are rejected or discarded as being spent, useless, worthless, or in excess. Table 8.1 provides a useful waste classification system. Solid waste is probably more correctly defined as any material thing that is no longer wanted. Defining solid waste is tricky, because solid waste is a series of paradoxes (O'Reilly, 1994):

- It is personal in the kitchen trash can but impersonal in a landfill.
- What one individual may deem worthless (an outgrown or out-of-fashion coat, for example) and fit only for the trash another individual can find valuable.
- It is of little cost concern to many Americans yet very costly to our society in the long term.
- It is an issue of serious federal concern yet a very localized problem from municipality to municipality.

The popular adage is accurate—everyone wants waste to be picked up, but no one wants it to be put down. It goes almost without saying that the other adage, "Not

---

TABLE 8.1
Classification of Solid Waste

Type	Principal Components
Trash	Highly combustible wastepaper, wood, cardboard cartons, including up to 10% treated papers, plastic, or rubber scraps; commercial and industrial sources
Rubbish	Combustible waste, paper, cartons, rags, wood scraps, combustible floor sweepings; domestic, commercial, and industrial sources
Refuse	Rubbish and garbage; residential sources
Garbage	Animal and vegetable wastes, restaurants, hotels, markets; institutional, commercial, and club sources

Source: Adapted from Davis & Cornwell (1991), *Introduction to Environmental Engineering*, 2nd ed., p. 585.

In My Back Yard" (NIMBY) is also accurate. The important point, though, is that whenever a material object is thrown away, regardless of its actual or potential value, it becomes a solid waste.

Garbage (with its tendency to decompose rapidly and create offensive odors and be commonly thrown almost anywhere and its plastic bags can be hazardous to our pets) is often used as a synonym for solid waste but actually refers strictly to animal or vegetable wastes resulting from handling, storage, preparation, or consumption of food.

The collective and continual production of all refuse (the sum of all solid wastes from all sources) is referred to as the *solid waste stream.* As stated previously, an estimated six billion metric tons of solid waste are produced in the United States each year. The two largest sources of solid wastes are agriculture (animal manure, crop residues, and other agricultural byproducts) and mining (dirt, waste rock, sand, and slag, the material separated from metals during the smelting process). About 10% of the total waste stream is generated by industrial activities (plastics, paper, fly ash, slag, scrap metal, and sludge or biosolids from treatment plants). The other 3% of the solid waste stream is made up of MSW, which is the focus of this chapter and consists of refuse generated by households, businesses, and institutions. Paper and paperboard account for the largest percentage (about 33%) of refuse materials by volume of MSW. Yard wastes are the next most abundant material, accounting for almost 13%. Glass and metals make up almost 18% of MSW, food wastes just under 13%, and plastics about 12.1%.

DID YOU KNOW?

USEPA (2007) points out that approximately 254 million tons of MSW were generated in the US in 2007, equivalent to a bit more than 0.8 pounds per person per day.

SOURCES OF MUNICIPAL SOLID WASTES

Sources of municipal solid wastes in a community are generally related to land use and zoning. MSW sources include residential, commercial, institutional, construction and demolition, municipal services, and treatment plants.

Residential Sources of MSW

Residential sources of MSW are generated by single and multifamily detached dwellings and apartment buildings. The types of solid wastes generated include food wastes, textiles, paper, cardboard, glass, wood, ashes, tin cans, aluminum, street leaves, and special bulky items including yard wastes collected separately, white goods (refrigerators, washers, dryers, etc.), batteries, oil, tires, and household hazardous wastes.

Commercial Sources of MSW

Commercial sources of MSW are generated in restaurants, hotels, stores, motels, service stations, repair shops, markets, office buildings, and print shops. The types of solid wastes generated include paper, cardboard, wood, plastics, glass, special wastes such as white goods and other bulky items, and hazardous wastes.

Institutional Sources of MSW

Institutional sources of MSW are generated in hospitals, schools, jails and prisons, and government centers. The types of solid wastes generated by institutional sources are the same as those generated by commercial sources.

Construction and Demolition Sources of MSW

Construction and demolition sources of MSW are generated at new construction sites, the razing of old buildings, road repair/renovation sites, and broken pavement. The types of solid wastes generated by construction and demolition sources are made up of standard construction materials such as wood, steel, plaster, concrete, and soil.

Municipal Services Sources of MSW

Municipal services (excluding treatment plants) sources of MSW are generated in street cleaning, landscaping, parks and beaches, recreational areas, and catch basin maintenance and cleaning activities. The types of solid wastes generated by municipal services are made up of rubbish, street sweepings, general wastes from parks, beaches, and recreational areas, and catch basin debris.

Treatment Plant Site Sources of MSW

Treatment plant site sources of MSW are generated in water, wastewater, and other industrial treatment processes (for example, incineration). The principal types of solid wastes generated in treatment plant sites are sludges or biosolids, fly ash, and general plant wastes.

LANDFILLING

Whether we use the term a *tip, dump, garbage dump, rubbish dump, dumping ground, midden* (*ancient pits for dumping waste*), or *landfill*, it is a site for disposal of waste materials. Landfilling our waste is nothing new; they are the oldest and most common form of waste disposal. Moreover, there is nothing new about the problems associated with landfilling our waste materials. Landfilling wastes has a history of causing environmental problems, including fires, explosions, production of toxic fumes, and storage problems when incompatible wastes are commingled. Landfills also have a history of contaminating surface and groundwaters. *Sanitary landfills* are designed and constructed to dispose of municipal solid wastes only. Not designed, constructed, or allowed to be operated for disposal of bulk liquids and/or hazardous wastes, landfills that can legally receive hazardous wastes are known as *secure landfills*.

Under RCRA, the design and operation of hazardous waste landfills have become much more technically sophisticated. Instead of the past practice of gouging out a huge maw from the subsurface and then dumping countless truckloads of assorted waste materials (including hazardous materials) into it until full, a hazardous waste landfill is now designed as a modular series of three-dimensional control cells. Design and operating procedures have evolved to include elaborate safeguards against leakage and migration of leachates. Secure landfills for hazardous waste disposal are equipped with double liners. Leakage detection, leachate collection and monitoring, and groundwater monitoring systems are required. Liners used in secure landfills must meet regulatory specifications. For example, the upper liner must consist of a 10–100-mil thick *flexible-membrane liner* (FML), usually made of sheets of rubber or plastic. The lower liner is usually FML, but recompacted clay of 3-ft thickness is also acceptable.

LEACHATE

Leachate is a landfill-produced liquid that in the course of passing through garbage (putrescible or industrial wastes) and other matter extracts, soluble or suspended solids, or any other compound of the material through which it passes. Many of these so-called leachate riders carry or entrain dissolved environmentally harmful substances that may then enter the environment. Simply stated and described, leachate can be described as any liquid material the drains from landfilled stockpile of material and contains significantly elevated concentrations of undesirable material derived from the material that it has passed through.

Note that landfill leachate varies widely in composition depending on the age of the landfill and the type of waste that it contains (Henry and Heinke, 1996; DOE, 1992). The generation of landfill leachate is generally caused by the precipitation percolating through the waste deposited in the landfill. Additional leachate volume is produced during the decomposition of carbonaceous material producing a wide range of other materials including methane, carbon dioxide, simple sugars, aldehydes, alcohols, and organic acids.

Secure landfills must be constructed to allow the collection of leachate (usually via perforated drainage pipes with an attached pumping system) that accumulates above each liner. Leachate control is critical. To aid in this control process (especially from leachate produced by precipitation), a low-permeability cap must be placed over completed cells. When the landfill is finally closed, a cap that will prevent leachate formation via precipitation must be put in place. This cap should be sloped to allow drainage away from the wastes. When a landfill is filled and capped, it cannot be completely abandoned, ignored, or forgotten. The site must be monitored to ensure that the leachate is not contaminating the groundwater. This is accomplished by installing downgradient test wells to assure detection of any leakage from the site.

When water percolates through waste, it promotes the blossoming of microbes, so to speak. A very distinct and diverse microbiome, essentially distinct from any other ecosystem that might be used for comparison such as sediments, soils, fresh water, salt water, permafrost, humans, bogs, and canines, confirms that physical and chemical factors influence the microbiome composition unique to landfills as compared to other environments. Numerous factors are related to the microbial composition among several separate groups within the landfill itself. These factors include the amount of chloride and barium, the rate of evapotranspiration, the age of waste, and the number of household chemicals present. These factors also affect the rate of degradation within the landfill.

Two of the microbes that have an enormous impact on the degradation of waste in landfills are bacteria and fungi. When water percolates through waste, decomposition by bacteria and fungi are enhanced. In turn, these processes release byproducts of decomposition and rapidly use up any available oxygen, creating an anoxic environment. While decomposition of waste is in progress, the temperature rises, and the pH falls rapidly with the result that many metal ions that are relatively insoluble at neutral pH become dissolved in the developing leachate. As the decomposition of waste material continues, more water is released adding to the volume of leachate. Leachate continues to grow and even begins to decompose materials, changing chemical compositions, that are not normally decomposable, such as cement-based building materials, fire ash, and gypsum-based materials. The deadly, rotten-egg-odor hydrogen sulfide (commonly called sewer gas) is present, in addition to other components of landfill gas, in those landfills containing large amounts of building waste such as gypsum. The physical appearance of the leachate when it exits a typical landfill is characteristically strongly smelling black-, yellow-, or orange-colored cloudy liquid. You may have heard the old saying that the smell would gag a maggot. Well, that accurately describes the smell of acidic leachate because of the presence of hydrogen-, nitrogen-, and sulfur-rich organic species called mercaptans (a colorless gas that smells like rotten cabbage).

When a landfill receives a mixture of municipal, commercial, and mixed industrial waste excluding significant amounts of concentrated chemical waste, the landfill leachate is characterized as a water-based solution of four groups of contaminants: dissolved organic matter (alcohols, acids, aldehydes, short-chain sugars), inorganic macro compounds (common cations and anions including sulfate, chloride, iron, aluminum, zinc, and ammonia), heavy metals (lead, nickel, copper, mercury), and xenobiotic organic compounds such as halogenated organics (PCBs, dioxins) (Kjeldsen et al., 2002). Note that a number of complex organic contaminants have also been detected in landfill leachates.

In addition to the contaminates listed and described above, USGS scientists found that pharmaceuticals, personal-care products, and other contaminants of emerging concern are widespread in water that has passed through landfills, known as leachate (USGS, 2021). USGS scientists studied 19 landfills across the United States and found 129 of 202 pharmaceutical (prescription and non-prescription), household, and industrial chemicals in untreated leachate samples (i.e., prior to treatment and environmental release). The number of chemicals measured in the leachate samples ranged from 6 to 82 (with a median of 31). An analysis of the data revealed that landfills located in areas receiving the greatest amounts of precipitation had the greatest number of chemicals detected and the highest concentration measured (USGS, 2021). Table 8.2 lists the chemicals and their maximum concentration that USGS found in their study.

LANDFILL GAS

In addition to the production of leachate, landfills also produce landfill gas (LFG). Harnessing the power of LFG is on the increase in the United States. Consider, for example, that as of 2020 there are 565 operational LFG energy projects in the United States and 477 landfills are being studied and appear to be good candidates for LFG projects (USEPA, 2020).

TABLE 8.2

Chemical Maximum Concentrations and Frequencies of Detection Observed (USGS, 2021) in Parts Per Trillion (ppt)

Maximum Concentration	Percent of Frequency Detection	Chemical
7,020,000 ppt	55	Para-cresol (plasticizer and flame-retardant, antioxidant in oils, rubber, polymers, and wood preservatives)
4,080,000 ppt	95	Bisphenol A (used in plastics, thermal paper, and epoxy resins)
705,000 ppt	65	Ibuprofen (analgesic, antipyretic)
254,000 ppt	95	DEET (insect repellent)
147,000 ppt	90	Lidocaine (local anesthetic, topical anti-itch treatment)
97,200 ppt	84	Camphor (natural compound with medicinal uses and embalming)
51,200 ppt	95	Cotinine (transformation product of Nicotine)
2,590 ppt	75	Carbamazepine (anticonvulsant and mood stabilizer)
168 ppt	55	Estrone (natural estrogenic hormone)

The natural decomposition of organic material is the source of LFG. LFG is composed of roughly 50% methane (the primary component of natural gas), 50% carbon dioxide (CO_2), and a small amount of other non-methane compounds. Note that methane is a potent greenhouse gas 28–36 times more effective than CO_2 at trapping heat in the atmosphere over a 100-year period, per the latest international climate change assessment reports.

So how does the methane emitted from landfills stack up against the other sources of methane? Emission estimates during 2018 are listed below:

- Natural Gas and Petroleum Systems—28%
- Enteric Fermentation—28% (digestive process by which carbohydrates are broken down by microorganisms into moles for absorption into the bloodstream of animals such as cattle, deer, and camels)
- MSW Landfills—15%
- Manure Management—10%
- Other—9%
- Coal Mining—8%
- Other Landfills—2%

As can be seen from the above list, MSW landfills are the third-largest source of human-related methane emissions in the United States, accounting for approximately 15.1% of these emissions in 2018. USEPA (2020) points out that the methane emissions from MSW landfills in 2018 were approximately equivalent to the greenhouse gas (GHG) emission from more than 20.6 million passenger vehicles driven for one year or the CO_2 emissions from more than 11.0 million homes' energy use for one year. At the same time, methane emissions from MSW landfills represent a lost opportunity to capture and use a significant energy resource.

It takes about a year before anaerobic conditions are established in an MSW landfill (see Figure 8.1). This is the case because when MSW is first deposited in the landfill, it undergoes an aerobic (with oxygen) decomposition stage when little methane is generated. After anaerobic conditions are established in the MSW landfill, bacteria (see Table 8.3) decompose the landfill organic waste in four phases. In Phase I, oxygen and nitrogen begin to decline and carbon dioxide and hydrogen begin to increase. In Phase II, hydrogen and carbon dioxide continue to increase and oxygen is depleted to zero. In Phase III, the anaerobic stage, carbon dioxide and nitrogen decrease and methane production increases. By Phase IV, carbon dioxide and methane are typically almost equal with methane usually holding a slight edge at 45–60% content.

Table 8.4 lists "typical" MSW landfill gases, their percent by volume, and their characteristics.

Conditions Affecting Landfill Gas Production

Research has shown that the rate and volume of landfill gas produced at a specific site depend on the characteristic of the waste (e.g., composition and age of the garbage) and a number of environmental factors (e.g., the presence of oxygen in the landfill, moisture content, and temperature) (ATSDR, 2001).

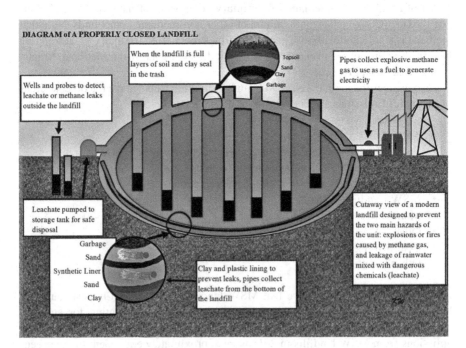

FIGURE 8.1 Cross-section of a municipal solid waste landfill. Source: USEPA, 2005, 2007; adapted by K. Welsh-Ware and F. Spellman from Aziz et al. (2018).

TABLE 8.3
The Four Phases of Bacterial Decomposition of Landfill Waste

In a landfill, bacteria decompose waste in four phases. The composition of the gas produced changes with each of the four phases of decomposition. Landfills often accept waste over a 20- to 30-year period, so waste in a landfill may be undergoing several phases of decomposition at once. This means that older waste in one area might be in a different phase of decomposition than more recently buried waste in another area.

Phase I

In the first phase of decomposition, aerobic bacteria—bacteria that live only in the presence of oxygen—consume oxygen while breaking down the long molecular chains of complex carbohydrates, proteins, and lipids that comprise organic waste. The primary byproduct of this process is carbon dioxide (see Figure 8.2). Nitrogen content is high at the beginning of this phase but declines as the landfill moves through the four phases. As long as oxygen is present, Phase I continues. Phase I decomposition can last for days or months, depending on how much oxygen is present when the waste is disposed of in the landfill. Oxygen levels vary according to factors such as how loose or compressed the waste was when it was buried.

Phase II

When oxygen in the MSW landfill is depleted, decomposition begins. Using an anaerobic process (a process that does not require oxygen), bacteria convert compounds created by aerobic bacteria into acetic, lactic, and formic acids, and alcohols such as methanol and ethanol. The landfill becomes highly acidic (see Figure 8.2). The acidic environment is important because as the acids mix with the moisture present in the landfill, they cause certain nutrients to dissolve, making nitrogen and phosphorus available to the increasingly diverse species of bacteria in the landfill. The gaseous byproducts of these processes are carbon dioxide and hydrogen. If the landfill is disturbed, or if oxygen is somehow introduced into the landfill, microbial processes will return to Phase I.

Phase III

When certain kinds of anaerobic bacteria consume the organic acids produced in Phase II and form acetate, (an organic acid; see Figure 8.2) Phase III decomposition begins. This causes the landfill to become a more neutral environment in which methane-producing bacteria begin to establish themselves. Methane- and acid-producing bacteria have a symbiotic, or mutually beneficial, relationship. Acid-producing bacteria create compounds for the methanogenic bacteria to consume. Methanogenic bacteria consume carbon dioxide and acetate, too much of which would be toxic to the acid-producing bacteria.

Phase IV

When both the composition and production rates of landfill gas (see Table 8.4) remain relatively constant, Phase IV decomposition begins. Phase IV landfill gas usually contains approximately 45–60% methane by volume, 40–60% carbon dioxide, and 2–9% other gases, such as sulfides. Gas is produced at a stable rate in Phase IV, typically for about 20 years; however, gas will continue to be emitted for 50 or more years after the waste is placed in the landfill (Crawford and Smith, 1985). Gas production might last longer, for example, if greater amounts of organics are present in the waste, such as at a landfill receiving higher than average amounts of domestic animal waste (ATSDR, 2001).

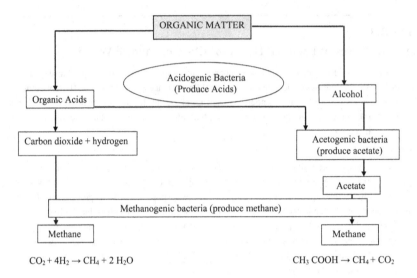

FIGURE 8.2   Organic matter to methane gas.

- *Waste composition*—the more organic waste present in a landfill, the more landfill gas (e.g., carbon dioxide, methane, nitrogen, and hydrogen sulfide) is produced by the bacteria during decomposition. The more chemicals disposed of in the landfill, the more likely non-methane organic compounds (NMOCs) and other gases will be produced either through volatilization or chemical reactions.
- *Age of garbage or refuse*—As a rule, more recently buried waste (i.e., waste buried less than 10 years ago) produces more landfill gas through bacterial decomposition, volatilization, and chemical reactions than does older waste (buried more than 10 years). Peak gas production usually occurs from 5 to 7 years after the waste is buried.
- *Presence of oxygen in the landfill*—only when oxygen is absent is methane produced.
- *Moisture content*—the presence of moisture (unsaturated conditions) in a landfill increases gas production because it encourages bacterial decomposition. Moisture may also promote chemical reactions that produce gases.
- *Temperature*—as the landfill's temperature rises, bacterial activity increases, resulting in increased gas production. Increased temperature may also increase rates of volatilization and chemical reactions.

MIGRATION OF LANDFILL GAS

Similar to the flow of electricity, landfill gases follow the path of least resistance. These landfill gases, once they are produced, tend to move away from the landfill. Gases tend to expand and fill the available space so that they move, or "migrate" through the limited pore spaces within the refuse and soils covering of the landfill. Because methane is lighter than air, it tends to move upward, usually through the

TABLE 8.4
Typical Landfill Gas Components

Component	Percent by Volume	Characteristics
Methane	45–60	Methane is a naturally occurring gas. It is colorless and odorless. Landfills are the single largest source of US man-made methane emissions.
Carbon dioxide	40–60	Carbon dioxide is naturally found at small concentrations in the atmosphere (0.03%). It is colorless, odorless, and slightly acidic.
Oxygen	0.1–1	Oxygen comprises approximately 21% of the atmosphere. It is odorless, tasteless, and colorless.
Ammonia	0.1–1	Ammonia is a colorless gas with a pungent odor.
Non-methane organic compounds (NMOCs)	0.01–0.6	NMOCs are organic compounds (i.e., compounds that contain carbon). (Methane is an organic compound but is not considered an NMOC.) NMOCs may occur naturally or formed by synthetic chemical processes. NMOCs most commonly found in landfills include acrylonitrile, benzene, 1.1-dichloroethane, 1,2-cis dichloroethylene, dichloromethane, carbonyl sulfide, ethylbenzene, hexane, methyl ethyl ketone, tetrachloroethylene, toluene, trichloroethylene, vinyl chloride, and xylenes.
Sulfide	0–1	Sulfides (e.g., hydrogen sulfide, dimethyl sulfide, mercaptans) are naturally occurring gases that give the landfill gas mixture its rotten-egg smell. Sulfides can cause unpleasant odors even at very low concentrations.
Hydrogen	0–0.2	Hydrogen is an odorless, colorless gas.
Carbon monoxide	0–0.2	Carbon monoxide is an odorless, colorless gas.

Source: EPA 2005; Tchobanoglous, Theisen, and Vigil 1993.

landfill surface. Upward movement of the landfill gas can be inhibited by densely compacted waste or landfill cover material (e.g., by daily soil cover and caps). (Note: Federal regulations require landfill operators to use at least six inches of earth material as a daily cover unless other materials are allowed as alternatives.) When upward movement is inhibited, the gas tends to migrate horizontally to other areas within the landfill or to areas outside the landfill, where it can resume its upward path. Some gases, such as carbon dioxide, are denser than air and will collect in subsurface areas, such as utility corridors. Three main factors influence the migration of landfill gases: diffusion (concentration), pressure, and permeability (ATSDR, 2001).

- Diffusion (concentration)—diffusion describes a gas's natural tendency to reach a uniform concentration in a given space, whether it is a room or the earth's atmosphere. Gases in a landfill move from areas of high gas concentrations to areas with lower gas concentrations. Because gas concentrations are generally higher in the landfill than in the surrounding areas, landfill gases diffuse out of the landfill to the surrounding areas with lower gas concentrations.

- Pressure—gases building up in a landfill create areas of high pressure in which the gas movement is restricted by compacted refuse or soil covers, and an area of low pressure in which gas movement is unrestricted. The variation in pressure throughout the landfill results in gases moving from areas of high pressure to areas of low pressure. The movement of gases from areas of high pressure to areas of lower pressure is known as convection. As more gases are generated, the pressure in the landfill increases, usually causing subsurface pressures in the landfill to be higher than either the atmospheric pressure or indoor air pressure. When the pressure in the landfill is higher, gases tend to move to ambient or indoor air.
- Permeability—Bases will also migrate according to where the pathways of least resistance occur. Permeability is a measure of how well gases and liquids flow through connected spaces or pores in refuse and soils. Dry, sandy soils are high permeable (many connected pore spaces), while moist clay tends to be much less permeable (fewer connected pore spaces). Gases tend to move through areas of high permeability (e.g., areas of sand or gravel) rather than through areas of low permeability (e.g., areas of clay or silt). Landfill covers are often made of low-permeability soils, such as clay. Gases in a covered landfill, therefore, may be more likely to move horizontally than vertically.

PUTTING LANDFILL GAS TO WORK

With LFG intended for multiple use, it is all about collecting, processing, and then using. Instead of allowing the gas to escape into the air, it can be captured, converted, and used as a renewable resource. There are collateral advantages to using LFG. For example, using LFG helps to reduce odors and other hazards associated with LFG emissions, and prevents methane from migrating into the atmosphere and contributing to local smog and global climate change. In addition, LFG energy projects generate revenue and create jobs in the community and beyond.

Before the landfill gas can be put to work, it must be collected. This is accomplished through vertical and horizontal piping systems buried within the MSW landfill. Sounds simple enough, but several factors, dealing mostly with design, must be considered before selecting the best technology option for an LFG recovery project. By and large, the volume of waste controls the potential amount of LFG that can be extracted from the landfill. Site conditions, LFG collection efficiency, and the flow rate for the extracted LFG also significantly influence the types of technologies and end-use options that are most feasible for a project. Design considerations for gas collection and treatment systems must be studied before constructing a landfill designed to be used for gas collection, treatment, and downstream use.

Gas Collection Systems

LFG collection systems can be configured as vertical wells, horizontal trenches, or a combination of both. Advantages and disadvantages of each type of well are listed in Table 8.5. Irrespective of whether wells or trenches are used, each wellhead is connected to lateral piping that transports the LFG to the main collection header, as shown in Figure 8.3. Note that it is important to design the collection system so that the operator can monitor and throttle (adjust) the gas flow, as necessary.

LFG Treatment Systems

Pure LFG gas is dirty; it must be treated to remove condensate, particulates, and other impurities. Treatment requirements depend on the end use.

TABLE 8.5

Advantages and Disadvantages of Vertical and Horizontal LFG Collection Wells (USEPA, 2020)

Vertical Wells		Horizontal Wells	
Advantages	Disadvantages	Advantages	Disadvantages
• Minimal disruption of landfill operations if placed in a closed area of the landfill • Most common design • Reliable and accessible for inspection and pumping	• Increased operation and maintenance required if installed in an active area of the landfill • Availability of appropriate equipment • Delayed gas collection if installed after site or cell closes	• Facilitates earlier collection of LFG • Reduced need for specialized construction equipment • Allows extraction of gas from beneath an active tipping area on a deeper site	• Increased likelihood of air intrusion will sufficiently be covered with waste • More prone to failure because of flooding or landfill settlement

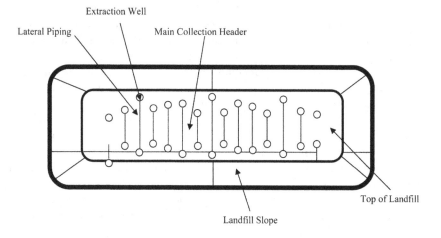

FIGURE 8.3 LFG extraction site plan.

- For LFG electricity projects, treatment systems usually include a series of filters to remove contaminants that can damage the components of the engine and turbine and reduce system efficiency.
- In the case of direct connection of LFG for use in boilers, furnaces, or kilns, minimal treatment is required.
- In order to comingle LFG with natural gas, advanced treatment is required to produce high-British thermal unit (Btu) gas.

Both primary and secondary treatment processing is used in treating LFG. Most primary processing systems include dewatering and filtration to remove moisture and particulates. In dewatering operations, the typical procedure is to employ "knockout" devices whereby free water and/or condensate are physically removed. Basically, a knockout device is a tank designed to perform a crude disengagement of gas and liquid. In operation, the area of the flow is increased and the velocity of the gas drops; therefore, the droplets or entrained liquid are no longer carried forward beyond this stage—achieving two ends with one treatment. However, it is common practice to use gas cooling and compression to remove water vapor or humidity from the LFG. Gas cooling and compression are proven techniques that have been used for many years and have become standard elements of active LFG collection systems. Secondary treatment systems are used to fine-tune the LFG, so to speak, to provide much greater gas cleaning than is possible using primary systems alone. Secondary systems usually employ both physical and chemical treatments. In the secondary treatment of LFG, it is all about intended end use. It is not uncommon to find siloxanes and sulfur compounds that need to be removed from LFG.

- *Siloxanes*—are found in solid waste (found in household and commercial products) and wastewater (i.e., sludge or biosolids within landfills). Siloxanes in the landfill volatize into LFG and are converted to silicon dioxide when the LFG is combusted. Silicon dioxide which is the main constituent of sand collects on the inside of internal combustion engines and gas turbines and on boiler tubes, increasing maintenance costs and reducing performance. Removal of siloxane can be both costly and challenging and the need for treatment depends on the level of siloxane in the LFG.
- Sulfides and disulfides (sulfur compounds) are corrosive in the presence of moisture—hydrogen sulfide is a good example in that its presence in concrete pipes corrodes and destroys the pipe over time.

Adsorption and absorption (or scrubbing) are the most common technologies used for the secondary treatment. Adsorption, which removes siloxanes from LFG, is a process by which contaminants adhere to the surface of an absorbent such as activated carbon or silica gel. Other gas treatment technologies that can remove siloxanes include subzero refrigeration and liquid scrubbing. Absorption removes compounds (such as sulfur) from LFG by introducing a solvent or solid reactant that produces a chemical/physical reaction. Advanced treatment technologies are used to remove carbon dioxide, non-methane organic compounds, and a wide variety of other contaminants in the LFG to produce a high-Btu gas (typically at least 96% methane).

DID YOU KNOW?

The main purpose of *carbon adsorption* used in advanced treatment processes is the removal of refractory organic compounds and soluble organic material that are difficult to eliminate by biological or physical/chemical treatment. In the carbon adsorption process, the waste stream passes through a container filled either with carbon powder or carbon slurry. Organics adsorb onto the carbon (i.e., organic molecules are attracted to the activated carbon surface and are held there) with sufficient contact time. A carbon system usually has several columns or basins used as contactors. Most contact chambers are either open concrete gravity-type systems or steel pressure containers applicable to either upflow or downflow operation. With use, carbon loses its adsorptive capacity. The carbon must then be regenerated or replaced with fresh carbon. Carbon used for adsorption may be in a granular form or powdered form.

Generating Electricity

The most common beneficial use application of using LFG is producing electricity, accounting for about three-fourths of all US LFG energy projects. The prime movers that use LFG to produce electricity are internal combustion engines, gas turbines, or microturbines.

The internal combustion engine is the most commonly used prime mover (conversion technology) in LFG applications because of its relatively low cost, high efficiency, and engine sizes that complement the gas output of many landfills. Internal combustion engines have been used at landfills where the gas quantity is capable of producing 800 to 3 MW, or where sustainable LFG flow rates to the engines are approximately 300–1,100 cfm at 50% methane. For projects larger than 3 MW, multiple engines can be combined together. Table 8.6 provides examples of available sizes of internal combustion engines.

TABLE 8.6
Internal Combustion Engine Sizes (USEPA, 2020)

Engine Size	Gas Flow (50% Methane)
540 kW	204 cfm
633 kW	234 cfm
800 kW	350 cfm
1.2 MW	500 cfm

cfm: cubic feet per minute kW: kilowatt MW: megawatt

A major benefit of internal combustion engines is that they are efficient as prime movers in converting LFG into electricity, achieving electrical efficiencies in the range of 30–40%. Even greater efficiencies are achieved where waste heat is recovered from the engine cooling system to make hot water, or for the engine exhaust to make low-pressure steam.

CASE STUDY 8.1 LYCOMING COUNTY LANDFILL COGENERATION

The success of Lycoming County landfill cogeneration is directly attributable to a creative permit and power purchase agreement structuring. Eighty percent of the Federal Bureau of Prisons' (FBOP) Allenwood Correctional Complex's electricity is supplied and 90% of the power and thermal needs of the Lycoming Country Landfill complex are met through the combustion of LFG in four internal combustion engines (6.2 MW total). The FBOP meets federal renewable energy requirements and gains long-term power price stability and cleaner energy, and the country receives funding for updating its gas collection system and revenue for the LFG without having to pay anything since PPL Renewable Energy develops, owns, and operates the project (USEPA, 2020).

In larger LFG energy projects, it is the gas turbine that is typically used. These are sites where LFG flows exceed a minimum of 1,300 CFM and are sufficient to generate a minimum of 3 MW. The main advantage of using gas turbines is their economies of scale. The cost per kW of generating capacity drops as the size of the gas turbine increases, and the electric generation efficiency generally improves as well. Simple-cycle gas turbines applicable to LFG energy projects typically achieve efficiencies of 20–28% at full load, yet these efficiencies drop substantially when the unit is running at partial load. System efficiency can be boosted to 40% when a combined-cycle configuration is used to recover the waste heat in the gas turbine exhaust to produce additional electricity. Again, like the simple-cycle gas turbines, combined-cycle configurations are less efficient at partial load.

The main advantages of gas turbines are that they are more resistant to corrosion damage than internal combustion engines and have lower nitrogen oxide emission rates. They are also relatively compact and have low operation and maintenance (O&M) costs compared with internal combustion engines. However, LFG treatment to remove siloxanes may be required to meet manufacturer specifications.

High-gas compression of 165 pound-force per square inch gauge (psig) or greater is required for gas turbines. This is their primary disadvantage because more of the plant's power is required to run the compression system.

Unlike the routine practice of feeding LFG from MSW landfills to internal combustion engines and gas turbines for the generation of electricity and for other varied purposes, microturbines have been used for landfill and biogas applications for a relatively short time, only since 2001. Note that costs for a microturbine project are higher than for the internal combustion engine project costs based on a dollar-per-kW installed capacity (Wang, Benson, Wheless, 2003). Notwithstanding the cost differential with using microturbines instead of internal combustion engines or gas turbines, there are several reasons for using microturbine technology:

* Require less LFG volume than internal combustion engines.
* Can use LFG with a lower percent methane (35% methane).
* Produce lower emissions of nitrogen oxides.
* Can add and remove microturbines as gas quantity changes.
* Interconnection is relatively easy because of the lower generation capacity.

Based on the operational experience related to microturbine operations, it has been found that early on in routine usage the microturbines were typically exposed to LFG that was not treated sufficiently; as a result, system failures were common. Typically, LFG treatment is required to remove moisture, siloxanes, and other contaminants. This treatment is composed of the use of inlet moisture separators, rotary vane-type compressors, coalescing filters, and adjustment to add 20–40°F above the dew point.

Sizes of 30, 70, and 250 kW are typical of microturbines using LFG. Projects should use the larger capacity microturbines where power requirements and LFG availability can support them. The use of a large microturbine can reduce capital cost and maintenance cost and increase efficiency.

CASE STUDY 8.2 LA CROSSE COUNT LANDFILL AND GUNDERSEN HEALTH SYSTEM CHP PROJECT, WISCONSIN

The Onalaska Campus is 100% energy independent, and Gundersen Health System is well on its way to meeting its total energy independence goal thanks to a public/private partnership with La Crosse County. A 2-mile pipeline brings LFG under Interstate 90 from the county landfill to create green power for the local grid and heat both buildings and water on the health system's campus. This combined heat and power (CHP) project serves as an excellent example of cost savings combined with environmental stewardship for other health systems nationwide that struggle with rising energy costs. The county benefits from a new revenue stream and its landfill is the first in the state to receive "Green Tier" status from the Wisconsin Department of Natural Resources.

Table 8.7 summarizes the advantages, disadvantages, and treatment requirements associated with each electricity-generating technology.

Pipe-to-Pipe Use of Medium-Btu Gas

The simplest and often the most cost-effective use of LFG is the direct pipe-to-pipe connection of medium-Btu fuel for boiler or industrial processes such as drying

TABLE 8.7

Advantages, Disadvantages and Treatment Requirements Summary (Electricity) (USEPA, 2020)

Advantages	Disadvantages	Treatment
Internal Combustion Engine		
• High efficiency compared with gas turbines and microturbines	• Relatively high maintenance	• At a minimum, requires primary treatment of LFG, for optimal engine performance, secondary treatment may be necessary
• Good size match with the gas output of many landfills	• Relatively high air emissions	
• Relatively low cost on a par kW installed capacity basis when compared with gas turbines and microturbines	• Economics may be marginal areas with low electricity costs	
• Efficiency increases when waste heat is recovered		
• Can add or remove engines to follow gas recovery trends		
Gas Turbine		
• Cost per kW of generating capacity drops as the size of the gas turbine increases, and the efficiency improves as well	• Efficiencies drop when the unit is running at partial load	• At a minimum, requires primary treatment of LFG, for optimal turbine performance, secondary treatment may be necessary
• Efficiency increases when heat is recovered	• Requires high gas compression	
• More resistant to corrosion damage	• High parasitic loads	
• Low nitrogen oxides emissions	• Economics may be marginal in areas with low electricity costs	
• Relatively compact	• Economics may be marginal in areas with low electricity costs	
Microturbine		
• Requires lower gas flow	• Economics may be marginal in Areas with low electricity costs	• Requires fairly extensive primary and secondary treatment of LFG
• Can function with lower percent methane		
• Low nitrogen oxide emissions		
• Relatively easy interconnection		
• Ability to add and remove units		

operations, kilns, and cement and asphalt production. In these projects, the LFG gas is piped directly to a nearby customer for use in combustion equipment as a replacement or supplementary fuel. The positive feature of this type of hookup and usage is that only limited condensate removal and filtration treatment are required, although some modifications of existing combustion equipment might be necessary.

In evaluating the sale of LFG in a pipe-to-pipe operation, the end-user's energy requirements are an important consideration. All gas that is recovered must be used as available, or it is essentially lost, along with associated revenue opportunities, because storing LFG is not economical. This enables the gas customer to have a steady annual gas demand compatible with the landfill's gas flow. When a landfill does not have an adequate gas flow to support the entire needs of a facility, LFG can still be used to supply a portion of the needs. For example, only one piece of equipment such as a boiler or set of burners is dedicated to burning LFG in some facilities.

DID YOU KNOW?

It may be possible to create a steady gas demand by serving multiple customers whose gas requirements are complementary. For example, an asphalt producer's summer gas load could be combined with a municipal building's winter heating load to create a year-round demand for LFG (USEPA, 2020).

CASE STUDY 8.3 HICKORY RIDGE LANDFILL AND COCA-COLA PROJECT, GEORGIA

Mas Energy developed a unique project that provides Coca-Cola's Atlanta Syrup Branch facility with a continuous supply of renewable electricity, steam, and chilled water. This is accomplished by conveying LFG via a 6-mile pipeline. The combined cooling, heat, and power (CCHP) system annually generates at least 48 million kilowatt-hours of onsite green power and provides nearly all the plant's energy needs, providing Coca-Cola real energy savings.

Table 8.8 provides the expected annual LFG flows from landfills of various sizes. Although the actual LFG flows will vary based on age, composition, moisture, and other factors of the waste, these numbers can be used as the first step toward assessing the compatibility of customer gas requirements and LFG output. A standard rule of thumb for comparing boiler fuel requirements with LFG output is that approximately 8,000–10,000 pounds per hour (lb/hr) of steam can be generated for every 1 million metric tons of waste-in-place at a landfill. So this means that a 5 million metric ton landfill can support the needs of a large facility requiring about 45,000 lb/hr of steam.

Note that the costs of equipment modifications and adjustments vary in accommodating the lower Btu value of LFG. Costs will be minimal if retuning the boiler burner is the only modification required. The costs associated with retrofitting

TABLE 8.8
Potential LFG Flows Based on Landfill Size (USEPA, 2020)

Landfill Size (Metric Tons Waste-in-Place)	Annual LFG Flow (MMBtu/yr)	Steam Flow Potential (lb/hr)
1,000,000	100,000	10,000
5,000,000	450,000	45,000
10,000,000	850,000	85,000

MMBtu/yr: Million British thermal units per year lb/hr: pounds per hour

boilers will vary from unit to unit depending on the boiler type, fuel use, and the age of the unit. Retrofitting boilers is typically required in the following situations:

- To ensure uninterruptible steam supply when retrofitting a unit incorporating LFG into a co-fired unit with other fuel where automatic controls are required to sustain a co-firing application. Also, to provide for immediate and seamless fuel switching in the event of a loss in LFG pressure to the unit. This retrofit will ensure an uninterruptible steam supply. Overall costs, including retrofit costs, (burner modifications, fuel train, and process controls), can range from $200,000 to $400,000.
- Modifying a unit that has a surplus or back-up steam supply so that the unit does not rely on the LFG to provide an uninterrupted supply of steam (a loss of LFG pressure can interrupt the steam supply). In this case, manual controls are implemented, and the boiler operating system is not integrated into an automatic control system. Overall costs can range from $100,000 to $200,000 (USEPA, 2020).

An option that can be employed to improve the quality of the gas to such a level that the boiler will not require a retrofit. Although the gas is not required to have a Btu value as high as pipeline-quality gas, it must be between medium- and high-Btu. This option eliminates the cost of a boiler retrofit and reduces maintenance costs for cleaning deposits associated with the use of medium-Btu LFG.

Again, it is important to note that a potential problem for boilers is the accumulation of siloxanes. The presence of siloxanes in the LFG causes a white substance to build up on the boiler tubes. Typically, this problem is managed by cleaning the boiler tubes. However, installing a gas treatment system to reduce the amount of siloxane in the LFG before it is delivered to the boiler is another option.

DID YOU KNOW?

LFG captured from the Lanchester Landfill in Narvon, Pennsylvania, is used for multiple purposes, including boilers, heaters, thermal oxidizers, ovens, engines, and turbines.

Along with using LFG to fuel some boiler systems, another ideal application is using LFG as a fuel for infrared heating in facilities that are close to a landfill. Infrared heating creates high-intensity energy that is safely absorbed by surfaces that warm up. In turn, these surfaces release heat into the atmosphere and raise the ambient temperature. Infrared heating applications for LFG have been successfully employed at several landfill sites in Europe, Canada, and the United States.

Infrared heaters are easy to install, require a small amount of LFG to operate, and are relatively inexpensive. Some operational projects (some of which have multiple heaters) use between 10 and 150 cfm. One positive aspect of using LFG for infrared heaters is that it does not require pretreatment unless siloxanes are present in the gas. Typically, one heater is required for every 500–800 square feet. Each heater costs approximately $3,000 and the cost of interior piping to connect the heaters with the building ceilings ranges from $20,000 to $30,000 (USEPA, 2020).

LFG can also supply heat for greenhouses, power grow lights, and heat water used in hydroponic plant cultures (see Figure 8.4). The costs for using LFG in greenhouses are highly dependent on how the LFG will be used. In the case of microturbine usage to power grow lights, the project costs would be similar to an equivalent microturbine LFG energy project. If the LFG is used to heat the green house, the cost incurred would be the cost of the piping and of the technology used, such as boilers.

Another example of the beneficial use of medium-Btu gas is in artisan studios with energy-intensive activities such as creating glass, metal, or pottery. In this case, it is not necessary to provide a large amount of LFG because it can be coupled with a commercial project. For instance, a gas flow of 100 cfm is sufficient for a studio that houses glassblowing, metalworking, or pottery kilns.

FIGURE 8.4 An example of hydroponic plants. Illustration by Kat Welsh-Ware and F. Spellman.

DID YOU KNOW?

The first artisan project to use LFG was at the EnergyXchange at the Yancey-Mitchell Landfill in North Carolina. LFG is used at this site to power two craft studios, four greenhouses, a gallery, and a visitor center (USEPA, 2020).

Another pipe-to-pipe direct application of medium-Btu gas is in using LFG for leachate evaporation. This is a good option for landfills where leachate disposal at a publicly owned treatment works (POTW) plant is unavailable or expensive. LFG is used to evaporate leachate to a more concentrated and more easily discarded effluent volume.

Evaporators are available in sizes to treat 10,000–30,000 gallons per day (gpd) of leachate. Capital costs range from $300,000 to $500,000. O&M costs range from $70,000 to $95,000 per year. When a system is owned and operated by a third party, long-term contracts will typically assess costs based on the volume of leachate evaporated. Some economies of scale are realized for larger-sized vessels, as indicated by a cost of leachate evaporation for a 10,000 gpd capacity unit at $0.10–$0.15 compared to a 30,000-capacity unit with a cost of about $0.05–$0.06 per gallon (USEPA, 2021).

It is interesting to note that the LFG pipe-to-pipe application can be used to heat boilers in plants that produce biofuels including biodiesel and ethanol. In this case, LFG is used directly as a fuel to offset another fossil fuel. Otherwise, LFG can be used as feedstock when it is converted to methanol for biodiesel production.

DID YOU KNOW?

An LFG biofuel project located in Sioux Falls, South Dakota, supplies LFG from the Sioux Falls Regional Sanitary Landfill to POET, a producer of biorefined products, for use in a wood waste-fired boiler, which generates steam for use in ethanol production (USEPA, 2020)

The advantages and disadvantages of pipe-to-pipe direct use of medium-Btu gas are summarized in Table 8.9.

LFG TO HIGH-BTU GAS

In accordance with the state regulations, LFG can be used to produce the equivalent of pipeline-quality gas (natural gas), compressed natural gas (CNG), or liquified natural gas (LNG). Pipeline-quality gas can be injected into a natural gas pipeline used for an industrial purpose. Otherwise, CNG and LNG can also be used to fuel vehicles at the landfill (such as water trucks, earthmoving equipment, light trucks, and autos), fuel refuse-hauling trucks (long-haul refuse transfer trailers and route collection trucks) and supply the general commercial market. Project costs depend

TABLE 8.9

LFG Direct-Use Advantages and Disadvantages and Treatment Requirements (USEPA, 2020)

Advantages	Disadvantages	Treatment
Boiler, Dryer and Kiln		
Uses maximum amount of recovered gas flow	Cost is tied to length of pipeline; Energy user must be nearby	Need to improve quality of gas or retrofit Equipment
Cost-effective		
Limited condensate removal and filtration treatment is required		
Does not require large amount of LFG and can be blended with other fuels		
Infrared Heater		
Relatively inexpensive	Seasonal use may limit LFG utilization	Limited condensate removal and filtration treatment
Easy to install		
Does not require a large amount of gas		
Can be coupled with another energy project		
Leachate Evaporation		
Good option for landfill where leachate disposal is expensive	High capital costs	Limited condensate removal and filtration treatment

on the purity of the gas required by the receiving pipeline or energy end user as well as the size of the project. Some economies of scale can be achieved when larger quantities of high-Btu gas can be produced.

DID YOU KNOW?

LFG can be converted into a high-Btu gas by increasing its methane content and, conversely, reducing its carbon dioxide, nitrogen, and oxygen content. In the United States, four methods have been commercially employed (beyond pilot testing) to remove carbon dioxide from LFG: water scrubbing, amine scrubbing, molecular sieve, and membrane separation.

Four Methods of Carbon Dioxide Removal from LFG

- *Water scrubbing* consists of a high-pressure biogas flow into a vessel column where carbon dioxide and some other impurities, including hydrogen sulfide, are removed by dilution in water that falls from the top of the vessel in the opposite direction of the gas flow. Methane is not removed because it has less dilution capability. The pressure is set at a point where only the carbon dioxide can be diluted; normally between 110 and 140 pounds per square inch (psi). The water that is used in the scrubbing process is then stripped in a separate vessel to be used again, making this system a closed loop that keeps water consumption low. The gases resulting from the stripping process (the same that were removed from the biogas) are then released or flared. Generally, no chemicals are required for the water scrubbing process, making it an attractive and popular technology.

Note: This technology will not remove certain contaminants such as oxygen and nitrogen that may be present in the raw biogas. This limitation may be an important variable when the end use of the cleaned gas is considered.

- *Amine scrubbing* usually utilizes amine as a physical solvent because it preferentially absorbs gases into the liquid phase in scrubbing systems designed to convert LFG to high-Btu gas. A typical Selexol-based plant employs the following steps:
- LFG compression (electric drive, LFG-fired engine drive, or product gas-fired engine drive).
- Moisture removal using refrigeration.
- Hydrogen sulfide removal in a solid media bed (using an iron sponge or a proprietary media).
- NMOC removal in a primary Selexol absorber.
- Carbon dioxide removal in a secondary Selexol absorber.

The LFG is placed in contact with the Selexol liquid in a Selexol absorber tower.

MNOCs are generally hundreds to thousands of times more soluble than methane. Also note that carbon dioxide is about 15 times more soluble than methane. Solubility also is enhanced with pressure, facilitating the separation of NMOCs and carbon dioxide from methane.

- *Molecular sieve* plants employ compression, moisture removal, and hydrogen sulfide removal steps, but rely on vapor-phase activated carbon to remove NMOC and a molecular sieve to remove carbon dioxide. Once exhausted, the activated carbon can be regenerated through a depressurizing heating and purge cycle. The molecular sieve process is also known as pressure swing adsorption.
- *Membrane separation* takes advantage of the physical property that gases will pass through polymeric membranes at differing rates. Carbon dioxide passes through the membrane approximately 20 times faster than methane. Pressure is the driving force for the separation process.

The primary cause of the presence of oxygen and nitrogen in LFG is air intrusion which can occur when air is drawn through the surface of the landfill and into the gas collection system. Air intrusion can often be minimized by adjusting well vacuums and repairing leaks in the landfill cover. In some instances, air intrusion can be managed by sending LFG from the interior wells directly to the high-Btu process and sending LFG from the perimeter wells (which often have higher nitrogen and oxygen levels) to another beneficial use of emissions control device. Although membrane separation can achieve some incidental oxygen removal, this is not the case for nitrogen; it is not removed. A molecular sieve can be configured to remove nitrogen by proper selection of media. Nitrogen removal, in addition to carbon dioxide removal, requires a two-stage molecular sieve pressure swing adsorption.

Compressed Natural Gas

For CNG production, the membrane separation and molecular sieve processes scale down more economically to smaller plants. For this reason, these technologies are more likely to be used for CNG production than the Selexol (amine scrubbing) process.

Liquefied Natural Gas

After LFG is first converted to CNG, LFG can be generated from the CNG. The CNG produced from LFG is liquified to produce LNG using conventional natural gas liquefaction technology. When assessing this technology, two factors should be considered:

- To avoid "icing" in the plant (carbon dioxide freezes at a temperature higher than methane liquefies), the CNG produced from LFG must have the lowest possible level of carbon dioxide. The low carbon dioxide requirement favors a molecular sieve over a membrane separation process or at least favors upgrading the gas produced by the membrane process with a molecular sieve. Water scrubbing also is an option.
- For those facilities that process large quantities of LNG, they are basically "design-to-order" facilities. A few manufacturers offer smaller, prepackaged liquefaction plants that have design capacities of 10,000 gpd or greater (USEPA, 2020).

Unless the nitrogen and oxygen content of the LFG is very low, additional steps must be taken to remove nitrogen and oxygen. Liquefier manufacturers desire inlet gas with less than 0.5 oxygen, citing explosion concerns. To produce LNG with a methane content of 96%, nitrogen needs to be limited.

Table 8.10 summarizes the advantages and disadvantages of converting LFG to high-Btu gas.

THE BOTTOM LINE

The slow, inexorable process of rock changing into soil gives us an ongoing demonstration of nature's unrelenting power. While we move around on the surface of the earth and push and shove soil into various stages and conditions, we forget one vital fact: we can do little to either slow or speed the natural process of soil creation.

TABLE 8.10

Advantages, Disadvantages and Treatment Requirements for High-Btu Gas

Advantages	Disadvantages	Treatment
Pipeline-Quality Gas		
Can be sold into a natural gas pipeline	Increased cost that results from tight management of wellfield operation needed to limit oxygen and nitrogen intrusion into LFG	Requires extensive and potentially expensive LFG processing
CNG or LNG		
Alternative fuels for vehicles at the landfill or refuse- hauling trucks, and for supply to the general commercial market	Increased cost that results from tight management of wellfield operation needed to limit oxygen and nitrogen intrusion into LFG	Requires extensive and potentially expensive LFG processing

Nature is a powerful and omnipresent force in our environments that we must both depend on and contend with. Human beings present another powerful force, and when you put their anthropogenic achievements into the equation (the heavy hand of humans and what we can do (and do) to our environment), the repercussions of both the destruction that Nature can dish out and the impact to the environment that humans contribute can become overwhelming to comprehend. Fortunately for our peace of mind, we don't often think along this line of reasoning. Is this a good thing? Is it good to ignore Nature and what she can do, what she does—at all times around us? Is it important to think about the impact humans have on their surroundings?

Many people simply don't care. They think,

We all know that humans impact their environment, but what the heck, the environment will take care of itself. And besides, why should we be concerned with what humans do today to the environment—soon it will all be forgotten? No matter what we do we leave our mark on the land.

We do impact the environment we use and inhabit, more than we are some-times aware. Have you ever walked through an old cemetery, one with crumbling and unreadable headstones? This can be a sobering reminder to all of us of our fragility—not only of human life but of the shortness of memory. The point: We must not forget; we must not only increase our awareness of our impact on our environment, but we must also increase our memory spans—our sense of history. For long-range planning, we need long-range vision—and the ability to remember what has gone on before.

With regard to solid waste, as we examine the problems associated with solid waste disposal, whether municipal, industrial, or hazardous, the answer to a question posed in this chapter becomes more and more apparent: When we "throw away" waste, it is not gone. Dealing with the waste permanently has only been postponed. Sometimes this postponement means that when we go back, the wastes are rendered helpful and harmless (as with some biodegradable wastes) but more often, it means that the problems we must face will be worse increased by chemistry and entropy. That 55-gallon drum was easier to handle before it rusted out.

Amid the cries of "not in my backyard" and "Pick up the Trash, but Don't Put It Down," we need to hear a more realistic and environmentally kinder truth: *There's no such thing as a free lunch.*

We pay, somehow, for what we get or use, whether we see the charges or not. The price for our solid waste habits will soon be charged to us. In some places (big cities, for example), the awareness of the size of the bill is sinking in.

Environmentally, what does that mean? In short, if we, as a society, are going to consume as we do, build as we do, grow as we do, we have to pay the price for our increase. And that, sometimes, is going to mean that our waste is going to be "in our backyard." We will have to increase the amount of solid waste we reuse and recycle; we will have to spend tax dollars to solve the problems with landfills and trash incineration, we will have to seriously look at how we live, how the goods we buy are packaged, how our industries deal with their wastes—because if we don't, the bill will be more than we can afford to pay.

REFERENCES AND RECOMMENDED READING

ASTM. (1969). *Manual on Water.* Philadelphia, PA: American Society for Testing and Materials.

ATSDR. (2001). *Landfill Gas Basics.* Atlanta, GA: Agency for Toxics Substances and Disease Registry.

Aziz, S.Q., Aziz, A.A., Bashir, M.O., and Mojiri, A. (2018). Statistical analysis of municipal solid waste landfill leachate. Accessed on 8 March 2021 @ https://www.epagov/landfills/municipal-solid-waste-landfills.

Beazley, J. (1992). *The Way Nature Works.* New York: Macmillan.

Blumberg, L., and Gottlieg, R. (1989). *War on Waste: Can America Win Its Battle with Garbage?* Washington, DC: Island Press.

Brady, N.C., and Weil, R.R. (1996). *The Nature and Properties of Soils*, 11th ed., Upper Saddle River, NJ: Prentice-Hall.

Carson, R. (1962). *Silent Spring.* Boston, MA: Houghton Mifflin Company.

Ciardi, J. (1997). From Stonework's, in *The Collected Poems of John Ciardi*, Cifelli, E. M. (ed.). Fayetteville, AR: University of Arkansas Press.

Crawford, J.F., and Smith, P.G. (1985). *Landfill Technology.* Oxford, UK: Butterworth-Heinemann.

Davis, G.H., and Pollock, G.L. (2003). Geology of Bryce Canyon National Park, Utah, in *Geology of Utah's Parks and Monuments*, 2nd ed., Sprinkle, D.A., et al. (eds.), Salt Lake City, UT: Utah Geological Association.

Davis, M.L., and Cornwell, D.A. (1991). *Introduction to Environmental Engineering.* New York: McGraw-Hill.

DOE Report. (1992). Solid waste landfill design manual. Accessed on 7 March 2021 @ http://usersox.ac.uk/~ayoung/landfill.html.

Environmental Justice. (1994). *Christian Science Monitor*, March 15.

Eswaran, H. (1993). Assessment of global resources: Current status and future needs. *Pedologie*, XL111: 19–39.

Foth, H.D. (1978). *Fundamentals of Soil Science*, 6th ed. New York: John Wiley and Sons.

Franck, I., and Brownstone, D. (1992). *The Green Encyclopedia*. New York: Prentice-Hall.

GLOBE Program. (2003). Teacher's guide: Soil investigation chapter. Accessed on 20 March 2009 @ http://archive.globe.gov/tctg/globetg.jsp.

Henry, J., and Heinke, G. (1996). *Environmental Science and Engineering*. New York: Prentice Hall.

Illich, I., Groeneveld, S., and Hoinacki, L. (1991). *Declaration on Soil*. Kassel, Germany: University of Kassel.

Kemmer, F.N. (1998). *Water: The Universal Solvent*. Oak Ridge, IL: NALCO Chemical Company.

Kjeldsen, P., Barlaz, M.A., Rooker, A.P., Baun, A., Ledin, A., and Christensen, R.H. (2002). Present and long-term composition of MSW landfill leachate: A review. *Critical Reviews in Environmental Science and Technology*, 32(4): 297–336.

Konigsburg, E.M. (1996). *The View from Saturday*. New York: Scholastic.

MacKay, D., Shiu, W.Y., and Ma, K.-C. (1997). *Illustrated Handbook of Physical-Chemical Properties and Environmental Fate for Organic Chemicals*. Boca Raton, FL: CRC Press/Lewis Publishers.

Morris, D. (1991). As if materials mattered. *The Amicus Journal*, 13(4): 17–21.

Mowet, F. (1957). *The Dog Who Wouldn't Be*. New York: Willow Books.

NPS. (2008). *The Hoodoo*. Washington, DC: National Park Service.

Oregon State University. (2003). Soil sampling for home gardens and small acreages. *Extension Service Bulletin*. Reprinted April 2003. Accessed on 20 March 2009 @ http://wwwagcomm.ads.orst.edu.

O'Reilly, J.T. (1994). *State & Local Government Solid Waste Management*. Deerfield, IL: Clark, Boardman, Callahan.

Peterson, C. (1987). Mounting garbage problem. *The Washington Post*, April 5.

Richards, B. (1988). Burning issue. *The Wall Street Journal*, June 16.

Soil Survey Staff, Keys to Soil Taxonomy. (1994). *Keys to Soil Taxonomy*. Washington, DC: USDA Natural Resources Conservation Service.

Spellman, F.R. (1998). *The Science of Environmental Pollution*. Boca Raton, FL: CRC Press.

Spellman, F.R. (2008). *The Science of Water*. Boca Raton, FL: CRC Press.

Spellman, F.R., and Whiting, N.E. (2006). *Environmental Science and Technology*, 2nd edition. Rockville, MD: Government Institutes.

Sposito, G. (2008). *The Chemistry of Soils*. New York: Oxford University Press.

Tchobanoglous, G., Theisen, H., and Vigil, S. (1993). *Integrated Solid Waste Management: Engineering Principles and Management Issues*. New York: McGraw-Hill.

Tomera, A.N. (1989). *Understanding Basic Ecological Concepts*. Portland, ME: J. Weston Walch, Publisher.

Tonge, P. (1987). All that trash. *Christian Science Monitor*, July 6.

USDA. (1975). *Soil Survey Staff, Soil Classification, A Comprehensive System*. Washington, DC: USDA Natural Resources Conservation Service.

USDA. (1999). *Soil Taxonomy: A Basic System of Soil Classification for Making and Interpreting Soil Surveys*, 2nd ed., Washington, DC: USDA Natural Resources Conservation Service.

USDA. (2009). *Urban Soil Primer*. Washington, DC: Untied States Department of Agriculture.

USEPA. (1988). *The Solid Waste Dilemma: An Agenda For Action—Background Document*. Washington, DC: EPA/530-SW-88-054A.

USEPA. (1989). *Decision-Makers Guide to Solid Waste Management*, Washington, DC: EPA/530-SW89-072.

USEPA. (1993). *Characterization of Municipal Solid Waste in U.S.: 1992 Update*. Washington, DC: EPA/530-5-92-019.

USEPA. (2005). *Guidance for Evaluating Landfill Gas Emissions from Closed or Abandoned Facilities.* Washington, DC: EPA-600/R-05/123a.

USEPA. (2007). Municipal solid waste generation, recycling, and disposal in the United States: Facts and figures for 2007. Accessed on 20 March 2009 @ www.epa.gov/osw.

USEPA. (2020). Basic information about landfill gas. Accessed on 7 March 2021 @ https://www.epa.gov/lmop/basic-information-about-landfill-gas.

USEPA. (2021). Benefits of landfill gas energy projects. Accessed on 13 March 2021 @ https://www.epa.gov/lmop/benefits-landfill-gas-energy-projects.

USGS. (2021). Pharmaceuticals and other chemicals common in landfill waste. Accessed on 3 March 2021 @ https://toxics.usgs.Gov/highlights/2014-08-12-leachate_pharm.html.

Wang, F., Smith, D.W., and El-Dimm, G. (2003). Scenario of landfilling in India. Accessed 6/02/2021 @ https://www.researchgate.net/publications.

Winegardner, D. (1996). *An Introduction to Soils for Environmental Professionals.* Boca Raton, FL: CRC Press.

Wolf, N., and Feldman, E. (1990). *Plastics: America's Packaging Dilemma.* Washington, DC: Island Press.

9 Dry Tombs to Wet Dumps

BIOREACTOR LANDFILLS

The natural production of landfill leachate occurs when rainwater filters through wastes placed in a landfill. When this liquid comes in contact with buried wastes, it leaches, or draws out, chemicals or constituents from those wastes—this process is Nature's Way. In order to work with Nature's Way in the production of landfills, leachate liquids are added to help bacteria break down the waste. The increase in waste degradation and stabilization is accomplished through the addition of liquid and air to enhance microbial processes. This bioreactor concept differs from the traditional "dry tomb" municipal landfill approach—the dry tomb morphs to the wet dump.

DID YOU KNOW?

A dry tomb landfill is a landfill that does not intentionally add liquids to the landfill to accelerate the decomposition process. Its main goal is to keep liquids out of the landfill.

Note: It is important to point out that USEPA and other agencies have been sponsoring and conducting various research and demonstration studies for bioreactor landfills since 1959. Most of the studies were completed in the early 1980s. The results of these studies showed that a landfill using a leachate recirculation design increased the rate of waste stabilization. This does not mean that we currently know all there is to know about bioreactors. It is more accurate to say that research continues, and knowledge gained is ongoing.

TYPES OF BIOREACTOR LANDFILLS

Currently, there are three types of bioreactor landfills: aerobic, anaerobic, and hybrid (USEPA, 2007).

- **Aerobic**—in an aerobic bioreactor landfill, leachate is removed from the bottom layer, piped to liquid storage tanks, and recirculated into the landfill in a controlled manner. Air is injected into the mass using vertical or horizontal wells to promote aerobic activity and accelerate waste stabilization.

DOI: 10.1201/9781003252665-9

- **Anaerobic**—in an anaerobic bioreactor landfill, moisture is added to the waste mass in the form of recirculated leachate and other sources to obtain optimal moisture levels. Biodegradation occurs in the absence of oxygen (anaerobically) and produces landfill gas. Landfill gas—primarily methane—can be captured to minimize greenhouse gas emissions and can be used for energy projects.
- **Hybrid (aerobic–anaerobic)**—the hybrid bioreactor landfill accelerates waste degradation by employing a sequential aerobic–anaerobic treatment to rapidly degrade organics in the upper section of the landfill and collect gas from lower sections. Operation as a hybrid results in the earlier onset of methanogenesis as compared to aerobic landfills.

In operation, the bioreactor landfill accelerates the decomposition and stabilization of waste. At a minimum, leachate is injected into the bioreactor to simulate the natural biodegradation process. Bioreactors often need other liquids such as storm water, wastewater, and wastewater treatment plant sludges to supplement leachate. This enhances the microbiological process by purposeful control of the moisture content and differs from a landfill that simply recirculates leachate for liquids management. Note that landfills that simply recirculate leachate may not necessarily operate as optimized bioreactors.

The most important factor promoting accelerated decomposition is moisture content. The bioreactor technology relies on maintaining optimal moisture content near field capacity—approximately 35–65%—and adds liquids when it is necessary to maintain that percentage. The moisture content, combined with the biological action of naturally occurring microbes, decomposes the waste. The microbes can be either aerobic or anaerobic. A side effect of the bioreactor is that it produces landfill gas (LFG) like methane in an anaerobic unit at an earlier stage in the landfill's life at an overall much higher rate of generation than traditional landfills.

Advantages of Bioreactor Landfills

There are several advantages of bioreactor landfills. Decomposition and biological stabilization of the waste in a bioreactor landfill can occur in a much shorter time than in a traditional "dry tomb" landfill. This can provide a potential decrease in long-term environmental risks and landfill and post-closure costs. Potential advantages of bioreactors include the following (USEPA, 2019):

- Decomposition and biological stabilization in years versus decades in "dry tombs"
- Lower waste toxicity and mobility due to both aerobic and anaerobic conditions
- Reduction in leachate costs
- A 15–30% gain in landfill space due to an increase in density of waste mass
- Significant increase in LFG generation that, when captured, can be used for energy use onsite or sold
- Reduction in post-closure care

Research has shown that due to the degradation of organics and the sequestration of inorganics, municipal waste can be rapidly degraded and made less hazardous by enhancing and controlling the moisture within the landfill under aerobic and/or anaerobic conditions. Leachate quality in a bioreactor rapidly improves, which leads to reduced leachate disposal costs. Landfill volume may also decrease with the recovered airspace, offering landfill operators the full operating life of the landfill.

Bioreactor-emitted LFG consists primarily of methane and carbon dioxide, as well as lesser amounts of volatile organic chemicals and/or hazardous air pollutants. Research indicates that the operation of a bioreactor may generate LFG earlier in the process and at a higher rate than the traditional landfill. The bioreactor LFG is also generated over a shorter period of time, because the LFG emissions decline as the accelerated decomposition process depletes the source waste faster than in a traditional landfill. The net result appears to be that the bioreactor produces more LFG overall than the traditional landfill does.

OPERATIONAL BIOREACTOR OVERVIEW

In 2007, USEPA selected five operational bioreactor sites in order to summarize their operations. The five sites selected were those that had quality data sets, high liquid addition rates, and were a mix of aerobic and anaerobic design. USEPA's evaluation and summary of the selected bioreactor sites focused on specific parameters affecting bioreactor performance, including liner head maintenance, settlement, side slope stability, fire prevention, and gas collection. Before summarizing USEPA's findings related to the specific parameters at the sites, the five sites that were selected and evaluated are listed and described as follows (USEPA, 2007):

- Crow Wing Country Landfill, Minnesota
- Williamson County Landfill, Tennessee
- Burlington County Landfill, New Jersey
- New River Regional Landfill, Florida
- Salem Country Landfill, New Jersey

A brief summary of the information pertinent to each site as a bioreactor facility is presented below.

CROW WING COUNTY LANDFILL

- Location—North Central Minnesota
- Owner—Crow Wing County
- Annual tonnage—50,000
- Permit method—leachate recirculation demonstration to current MSW landfill permit
- Extent and area—full scale; 14.1 acres
- Type—leachate recirculation; anaerobic
- Year started—1998

- Method of injection—Treated and untreated leachate is injected via horizontal laterals,
- Annual volume recirculated: 4 million gallons

WILLIAMSON COUNTY LANDFILL

This bioreactor was the only full aerobic bioreactor of the five listed herein:

- Location—Central Tennessee, Williamson County
- Owner—Williamson County
- Annual tonnage—about 70,000 tons disposed in a closed landfill on approximately 7 acres
- Permit method—state permit for leachate recirculation
- Extent and area—full scale in 7-acre closed landfill
- Type—leachate and stormwater recirculation; aerobic
- Year started—June 2000
- Method of injection—leachate and air are injected into vertical risers with force main and header from the storage tank that was retrofitted for the closed landfill
- Annual volume recirculated—about 1 million gallons

BURLINGTON COUNTY BIOREACTOR

- Location—Northwest New Jersey
- Owner—Burlington County
- Tonnage in-place—about 1 million tons of MSW in a 10-acre area
- Permit method—New Jersey department of environmental protection (NJDEP) permit for leachate recirculation including leachate from closed landfill, stormwater runoff from co-compost area, stormwater ponds, grey water and sewage from office and lab complex, and surface water
- Extent—full-scale leachate and liquid recirculation installed as landfill was being built
- Type—anaerobic recirculation from 2002 to 2005
- Method of injection—leachate recirculated into horizontal pipes and trenches with force main connection to storage tank
- Volume recirculated—about 18 million gallons to date

NEW RIVER REGIONAL BIOREACTOR

This bioreactor is the only "research" bioreactor designed, operated, and permitted with State of Florida grant funding.

- Location—North Central Florida
- Owner—Union County
- Tonnage in-place—about 1 million tons in an existing filled 10-acre area

- Permit method—Florida dept of environmental protection (FDEP) permit for leachate recirculation
- Extent—full-scale leachate recirculation into an existing interim capped landfill with an exposed membrane cover
- Type—anaerobic bioreactor in 75% of sit and aerobic bioreactor in 25% of landfill boundary
- Method of injection—air and leachate injected in nested vertical risers
- Volume recirculated—about 6.5 million gallons injected to date

SALEM COUNTY BIOREACTOR

- Location—Southwest New Jersey
- Owner—Salem County
- Permit method—NJDEP permit for leachate recirculation
- Extent and area—full-scale leachate recirculation as landfill was constructed; over 5-acre area
- Type—anaerobic bioreactor with leachate recirculation since 2000 and MSW moisture added to date is about 44 gallons/ton
- Method of injection—leachate recirculated from the storage tank and force main to subsurface horizontal injection trenches

SUMMARY OF FINDINGS

Liner head maintenance: The landfill that measured the highest leachate loading rate (recirculating rate) was Crow Wing County. In 2005, about 75 gallons of leachate per ton of MSW were recirculated. There have been no problems noted in maintaining less than or equal to 30 cm of the head of leachate on the primary liner. Crow Wing Country Landfill also is the only landfill of the five evaluated that has a leak detection system (lysimeters) below the primary liner and leachate sump. The flow in the leak detection system was found to be below levels of concern and did not correlate with the rates of leachate recirculation or leachate generation and/or rainfall over the history of the landfill.

The other four bioreactor landfills had no historical problems in maintaining less than 30 cm (1 ft) of the head of leachate on the liner. This is in accordance with 40 CFR § 264.301 Landfill Design and Operating Requirements (a) (2), which states:

> A leachate collection and removal system immediately above the liner that is designed, constructed, maintained, and operated to collect and remove leachate from the landfill. The Regional Administrator will specify design and operating conditions in the permit to ensure that the leachate depth over the liner does not exceed 30 cm.

(Note that the drainage length, drainage slope, permeability of the drainage materials, and leachate impingement rate control the head on the liner.) Williamson County Landfill and New River Regional Landfill reported historically low leachate head (10 cm or less) during the bioreactor operations. New River Regional Landfill also was the only bioreactor of the five evaluated that was able to maintain low heads with a "pipe-less" collection system design using only tri-planar geocomposite for the drainage media with gravity flow to a collection sump. There was also no apparent correlation of leachate recirculation rates, leachate generation rates, and liner head maintenance in any of the five bioreactors reviewed. This most likely is due to good moisture distribution of the leachate recirculation system designs and operations that evenly dispersed leachate laterally and vertically into the waste mass to the point of absorption (i.e., less than field capacity).

As was demonstrated at each site, with each having a different leachate collection approach, the engineered systems were all functioning as intended to maintain head less than 1 ft over the liner.

Settlement: USEPA found that a common variable between all five of the bioreactor landfills was the use and recommendation of high density polyethylene pipe (HDPT) pipe (solid and perforated) flexible enough to handle settlement. Experience at Williamson County Landfill bioreactor has shown that the type of piping material selected is important to the delivery of air and liquids. At first, PVC header piper and joint connections were used. This was found to be brittle after exposure to the sun and also was subject to settlement and cracking. As a result, air leaks were found (which was important since this was an aerobic bioreactor). After piping was replaced with HDPE, there have been no problems with integrity even with an additional settlement of the landfill.

The only occurrence of problems with infrastructure and settlement was reported by the operators of the Burlington County Landfill bioreactor. They observed that the lateral injection and gas collection system piping was pulled inward to the landfill slope with the settlement, causing the header piper connection to crimp or kink. This problem later was corrected by the replacement of a new pipe with adequate "slack" within it, especially at the connection of header and lateral piping.

The New River Regional Landfill also allowed for settlement in their design and installation and did not observe any structural damage to infrastructure. The Salem County Landfill bioreactor also used and recommended HDPE pipe with extra slack to allow for settlement and has not noted any infrastructure problems due to settlement. Crow Wing Country Landfill bioreactor used a unique design of 4″ pipe place in an overlapping 5″ pipe with a Fernco flexible coupling. This allowed sufficient flexibility in the joint and sack for settlement. Since 1998 when leachate recirculation started, no problems with leachate injection systems have been observed.

Side slope stability: The Williamson County aerobic bioreactor is the only landfill that witnessed side slope issues. This facility had two minor veneer failures, most likely due to its steep side slope of 1.5:1. The first failure was in a small area and was only the compost over slumping due to heavy rainfall and runoff. There was no failure noted in the waste mass within the landfill or due to bioreactor operations. The second failure was due to excess pressure noted in one well which may have been created by air and leachate injection. This veneer slumpage appears to have been caused by leachate seeps midway up the slope. After the cover was replaced, there were no further incidents, as leachate and air quantities were lowered for injection back into this section of the landfill.

Burlington County Landfill had leachate seeps most likely due to gas wells that leaked water out and leachate recirculation lines that were perforated within 7.6 in (25 ft) of the side slope. This was corrected and has never occurred again, as gas wells were equipped with dewatering pumps and leachate recirculation was terminated in this section of the landfill. None of the other landfills had side slope issues or instability (USEPA, 2007)

Experience has shown that bioreactors that design and operate with the prevention of potential leachate seeps in sides slopes do not experience slope instability. A functioning leachate injection line from 15 to 30 m (50–100 ft) from the edge of the outside slope should ensure that pore pressures will not build up to affect slopes. Moreover, experience has demonstrated that the use of alternate and permeable daily cover will help avoid seeps and instability.

Fire prevention: Whenever methane and combustible material are present in certain quantities, the danger of fire and explosion are real. The only item missing for a methane- and combustibles-fire is an ignition source. MSW landfills are methane producers. Two of the five landfills reported experiencing conditions which posed the potential for "fires" (USEPA, 2007). Williamson County Landfill bioreactor had a "hot spot" over their temperature goal which was quickly remedied by adding more leachate and reducing air injection. No hot spots occurred again as monitoring and liquid management controlled the temperatures. Burlington County Landfill had a "hot spot" that appeared to be a subsurface fire but was readily controlled by reducing vacuum on the gas wells and balancing the system. The condition was caused by too much air withdrawn through the thin interim permeable cover. It was controlled by the addition of extra leachate injected in the area.

It appears that fire prevention and occurrence are similar in frequency to "dry tomb" landfills and are a matter of gas tuning, balancing, and monitoring. Excessive air intrusion and dryness of MSW contribute to fires in any landfill. Aerobic bioreactor landfills may be the most vulnerable, but it is a matter of monitoring and control of liquid addition. Experience has shown that in aerobic landfills the simultaneous addition of liquids and reduction in air injection volume will help control temperatures in a desirable operating range.

Gas collection: No gas collection systems were necessary for the non-aerobic landfills that included Williamson County Landfill and the aerobic portion of the New River Regional Landfill. Methane that existed in both of these "retro-fit" landfills was oxidized within a day or two of operation and did not pose a threat to the environment as there were no gas migration issues. Also, due to its size, Crow Wing County Landfill did not install active gas systems as it is passively vented. Plans are to install an active system within the next year and sell landfill gas for energy.

The other anaerobic bioreactors that had installed gas collection systems designed and installed them during the construction of the leachate injection systems. Both horizontal and vertical gas collection systems were installed and operated. A common theme was to install dewatering pumps and also to locate gas extraction systems far away from active leachate injection systems. Moreover, some landfills did not operate gas collection in areas of active leachate injection. Most sites also rotated the injection of leachate around the site so as to not over saturate any one area and to allow time for leachate to drain. This should also provide relief to gas collection systems.

Note that the Burlington County Landfill reactor operators found that if vertical gas wells are installed deep within the zone of active biodegradation, then the temperatures in the waste are such that PVC pipe will weaken. There were a few vertical gas wells that were "crimped" at depth. They will be replaced with chlorinated polyvinyl chloride (CPVC) that has a higher melting (or weakening) point.

THE BOTTOM LINE

The five selected full-scale operating bioreactor landfills demonstrate that these types of landfills can comply with pertinent federal/state/local regulations and technical guidance. The addition of leachate and other liquids can be managed with the appropriate design and operation of injection systems that evenly distribute the moisture within the waste. The design of the leachate collection systems appears to be adequate to handle any additional leachate generated as all sites have been able to maintain leachate levels under 30 cm of head on the line. Slope stability issues have been minor and are readily corrected. Proper design and operations also can provide for slope stability. Fires or "hot spots" appear to have greater potential in aerobic landfills but can be managed with good monitoring and prompt addition of liquids. The anaerobic bioreactors have similar issues as normal landfill-balancing, tuning, and monitoring of the gas extraction systems (USEPA, 2007).

REFERENCES

USEPA. (2007). *Bioreactor Performance*. Washington, DC: United States Environmental Protection Agency—EPA530-R-07-007.
USEPA. (2019). *Bioreactors Fact Sheet*. Washington, DC: United States Environmental Protection Agency. Accessed 03/03/2021 @ https://www.epa.gov/sites/production/files/2019-08/documents/membrane_bioreactor_fact_sheet_p100:/zg.pdg.

10 Incineration

WASTE INCINERATION

Simply, an incinerator is a furnace for waste disposal via the combustion and burning process. Not surprisingly, incinerators produce the maximum solids and moisture reductions. They incinerate organic substances contained in waste materials. The endpoint of high-temperature thermal treatment of waste produces ash, flue gas, and heat—waste material. For incineration of wastes, the ancillary equipment required depends on whether the unit is a multiple hearth or fluid-bed incinerator or possibly either a rotary kiln or moving grate. In this work, because the multiple hearth moving grate and fluid-bed type incinerators are most commonly used to incinerate waste, this discussion focuses on these two types. In regard to ancillary equipment generally required to support the incinerator, each type of system requires a source of heat to reach ignition temperature, a waste feed system, and ash handling equipment. The ash is formed by the inorganic constituents of the waste and may take the form of particulates or solids carried by the flue gas. It is important to note that the system must also include all required equipment (e.g., scrubbers) to achieve compliance with air pollution control requirements—the flue gases must be cleaned (flue gas cleaning) of gaseous and particulate pollutants before they are dispersed into the atmosphere.

Solid wastes are conveyed or pumped to the incinerator. The solids are dried and then ignited (burned). As they burn, the organic matter is converted to carbon dioxide and water vapor and the inorganic matter is left behind as ash or "fixed" solids. The ash is then collected for reuse or disposal.

Note: In the following discussion, the focus is on the incineration of sewage sludge or biosolids; however, keep in mind that waste incineration is about the same no matter the feedstock, including municipal solid waste (MSW), and the end result of flue gas, ash, and slag are about the same, no matter the type of waste.

SLUDGE INCINERATION PROCESS DESCRIPTION

The incineration process first dries then burns the waste. Generally, the process involves the following steps:

(1) The temperature of the sludge feed is raised to 212°F.
(2) Water evaporates from the sludge.
(3) The temperature of the water vapor and air mixture increases.
(4) The temperature of the dried sludge volatile solids raises to the ignition point.

Note: Incineration will achieve maximum reductions if sufficient fuel, air, time, temperature, and turbulence are provided.

Incineration Processes

Multiple Hearth Furnace

The *multiple hearth furnace* consists of a circular steel shell surrounding a number of hearths. Scrappers (rabble arms) are connected to a central rotating shaft. Units range from 4.5 to 21.5 ft in diameter and have from 4 to 11 hearths. In operation, dewatered sludge solids are placed on the outer edge of the top hearth. The rotating rabble arms move them slowly to the center of the hearth. At the center of the hearth, the solids fall through ports to the second level. The process is then repeated in the opposite direction. Hot gases are generated by burning on lower hearths dry solids. The dry solids pass to the lower hearths. The high temperature on the lower hearths ignites the solids. Burning continues to completion. Ash materials discharge to lower cooling hearths where they are discharged for disposal. Air flowing inside the center column and rabble arms continuously cools internal equipment.

Fluidized Bed Furnace

The *fluidized bed* incinerator consists of a vertical circular steel shell (reactor) with a grid to support a sand bed, and an air system to provide warm air to the bottom of the sand bed. The evaporation and incineration process takes place within the superheated sand bed layer. In operation, the air is pumped to the bottom of the unit. The airflow expands (fluidize) the sand bed inside. The fluidized bed is heated to its operating temperature (1,200–1,500°F). Auxiliary fuel is added when needed to maintain operating temperature. The sludge solids are injected into the heated sand bed. Moisture immediately evaporates. Organic matter ignites and reduces to ash. Residues are ground to fine ash by the sand movement. Fine ash particles flow up and out of the unit with exhaust gases. Ash particles are removed using common air pollution control processes. Oxygen analyzers in the exhaust gas stack control the airflow rate.

Note: Because these systems retain a high amount of heat in the sand, the system can be operated as little as four hours per day with little or no reheating.

Operational Observations, Problems, and Troubleshooting

The operator of an incinerator monitors various performance factors to ensure optimal operation. These performance factors include feed sludge volatile content, feed sludge moisture content, operating temperature, sludge feed rate, fuel feed rate, and air feed rate.

Note: To ensure that the volatile material is ignited, the sludge must be heated between 1,400 and 1,700°F.

To be sure that operating parameters are in the correct range, the operator monitors and adjusts sludge feed rate, airflow, and auxiliary fuel feed rate. All maintenance conducted on an incinerator should be in accordance with the manufacturer's recommendations.

Operational Problems

The operator of a multiple hearth or fluidized bed incinerator must be able to recognize operational problems using various indicators or through observations. We discuss these indicators/observations, causal factors, and recommended corrective actions in the following sections to make the point that operating a sludge/biosolids or MSW incinerator is more complex than simply opening the trash deposit door and dumping the sludge or trash.

1. **Multiple Hearth: Incinerator Temperature Too High**
   Causal Factors:

   - Excessive fuel feed rate.
   - Greasy solids.
   - Thermocouple burned out.

   Corrective Actions (where applicable):

   - Decrease fuel feed rate.
   - Reduce sludge feed rate.
   - Increase air feed rate.
   - Replace thermocouple.

2. **Multiple Hearth: Furnace Temperature Too Low**
   Causal Factors:

   - Moisture content of the sludge has increased.
   - Fuel system malfunction.
   - Excessive air feed rate.
   - Flame out.

   Corrective Actions (where applicable):

   - Increase fuel feed rate until de-watering operation improves.
   - Establish proper fuel feed rate.
   - Decrease air feed rate.
   - Increase sludge feed rate.
   - Relight furnace.

3. **Multiple Hearths: Oxygen Content of Stack Gas Too High**
   Causal Factors:

   - Sludge feed rate is too low.
   - Sludge feed system blockage.
   - Air feed rate is too high.

   Corrective Actions (where applicable):

   - Increase sludge feed rate.
   - Clear any feed system blockages.
   - Decrease air feed rate.

4. **Multiple Hearths: Oxygen Content of Stack Gas Too Low**
   Causal Factors:

   - Volatile or grease content of the sludge has increased.
   - Air feed rate is too low.

   Corrective Actions (where applicable):

   - Increase air feed rate.
   - Decrease sludge feed rate.
   - Increase air feed rate.

5. **Multiple Hearths: Furnace Refractories Deteriorated**
   Causal Factor:

   - Rapid startup/shutdown of the furnace.

   Corrective Actions:

   - Repair furnace refractories.
   - Follow specified startup/shutdown procedures.

6. **Multiple Hearths: Unusually High Cooling Effect**
   Causal Factor:

   - Air leak.

   Corrective Action:

   - Locate and repair leak.

7. **Multiple Hearths: Short Hearth Life**
   Causal Factor:

   - Uneven firing.

   Corrective Action:

   - Fire hearths equally on both sides.

8. **Multiple Hearths: Center Shaft Shear Pin Failure**
   Causal Factors:

   - Rabble arm is dragging on hearth.
   - Debris is caught under the arm.

   Corrective Actions (where applicable):

   - Adjust rabble arm to eliminate rubbing.
   - Remove debris.

9. **Multiple Hearth: Scrubber Temperature Too High**
   Causal Factor:

   - Low water flow to the scrubber.

   Corrective Action:

   - Adjust water flow to proper level.

10. **Multiple Hearth: Stack Gas Temperatures Too Low**
    Causal Factors:

    - Inadequate fuel feed supply.
    - Excessive sludge feed rate.

Corrective Actions (where applicable):
- Increase fuel feed rate.
- Decrease sludge feed rate.

11. **Multiple Hearth: Stack Gas Temperatures Too High**
Causal Factors:

- Sludge has higher volatile content (heat value).
- Excessive fuel feed rate.

Corrective Actions (where applicable):

- Increase air feed rate.
- Decrease sludge feed rate.
- Decrease fuel feed rate.

12. **Multiple Hearth: Furnace Burners Slagging Up**
Causal Factor:

- Burner design.

Corrective Action:

- Replace burners with newer designs that reduce slagging.

13. **Multiple Hearths: Rabble Arms Dropping**
Causal Factors:

- Excessive hearth temperatures.
- Loss of cooling air.

Corrective Actions (where applicable):
- Maintain temperatures within proper range.
- Discontinue injection of scum into the hearth.
- Repair cooling air system immediately.

14. **Multiple Hearths: Excessive Air Pollutants in Stack Gas**
Causal Factors:

- Incomplete combustion—insufficient air.
- Air pollution control malfunction.

Corrective Actions (where applicable):

- Raise air to fuel ration.
- Repair/replace broken equipment.

15. **Multiple Hearths: Flashing or Explosions**
Causal Factor:

- Scum or grease additions.

Corrective Action:

- Remove scum/grease before incineration.

16. **Fluidized Bed: Bed Temperature Falling**
Causal Factors:

- Inadequate fuel supply.
- Excessive sludge feed rate.

- Excessive sludge moisture levels.
- Excessive air flow.

Corrective Actions (where applicable):

- Increase fuel supply.
- Repair fuel system malfunction.
- Decrease sludge feed rate.
- Correct sludge de-watering process problem.
- Decrease airflow rate.

17. **Fluidized Bed: Low (<3%) Oxygen in Exhaust Gas**
Causal Factors:

- Low air flow rate.
- Fuel feed rate too high.

Corrective Actions (where applicable):

- Increase blower air feed rate.
- Reduce fuel feed rate.

18. **Fluidized Bed: Excessive (>6%) Oxygen in Exhaust Gas**
Causal Factor:

- Sludge feed rate is too low.

Corrective Actions (where applicable):

- Increase sludge feed rate.
- Adjust fuel feed rate to maintain steady bed temperature.

19. **Fluidized Bed: Erratic Bed Depth on Control Panel**
Causal Factor:

- Bed pressure taps plugged with solids.

Corrective Actions (where applicable):

- Tap a metal rod into the pressure tap pipe when the unit is not in operation.
- Apply compressed air to pressure tap while the unit is in operation (follow manufacturer's safety guidelines).

20. **Fluidized Bed: Preheat Burner Fails and Alarm Sounds**
Causal Factors:

- Pilot flame is not receiving fuel.
- Pilot flame is not receiving spark.
- Defective pressure regulator(s).
- Pilot flame ignites but flame scanner malfunctions.

Corrective Actions (where applicable):

- Correct fuel system problem.
- Replace defective part.
- Replace defective regulator(s).
- Clear scanner sight glass.
- Replace defective scanner.

21. **Fluidized Bed: Bed Temperature Too High**
Causal Factors:

- Bed gun fuel feed rate is too high.
- Grease or high organic content in sludge (high heat value).

Corrective Actions (where applicable):

- Reduce bed gun fuel feed rate.
- Increase airflow rate.
- Decrease sludge fuel rate.

22. **Fluidized Bed: Bed Temperature Reads Off Scale**
Causal Factor:

- Thermocouple burned out.

Corrective Action:

- Replace thermocouple.

23. **Fluidized Bed: Scrubber Inlet Shows High Temperature**
Causal Factors:

- No water flowing in the scrubber.
- Spray nozzles are plugged.
- Ash water not recirculating.

Corrective Actions (where applicable):

- Open valves to provide water.
- Correct system malfunction to provide required pressure.
- Clear nozzles and strainers.
- Repair/replace recirculation pump.
- Unclog scrubber discharge line.

24. **Fluidized Bed: Poor Bed Fluidization**
Causal Factor:

- Sand leakage through support plate during shut down.

Corrective Actions (where applicable):

- Clear wind box.
- Clean wind box at least once per month.

DID YOU KNOW?

Currently, there are 75 facilities in the United States that recover energy from the combustion of municipal solid waste. These facilities exist in 25 states, mainly in the Northeast.

BENEFICIAL REUSE OF INCINERATED MSW

When MSW is incinerated, the original waste is transformed into flue gas and ash. The flue gas is usually handled using air pollution technology such as scrubbers to neutralize the harmful effects of the gas in the atmosphere. With the leftover ash, it is a different story. It has to be removed from the incinerator and put somewhere, usually in some landfill or other dumpsite. However, scrubbing incinerator exhaust and handling ash can be accomplished in a manner that is beneficial.

WASTE TO ENERGY

In the case of turning the flue gas generated by incineration of MSW, the exhaust gas can be used as an energy source via energy recovery technology (e.g., combustion, gasification, pyrolization, anaerobic digestion, and landfill gas recovery) to produce usable heat, electricity, or fuel. This process is called waste to energy.

Energy recovery from the incineration of MSW is a key part of USEPA's non-hazardous waste management hierarchy (see Figure 10.1), which ranks various management strategies from most to least environmentally preferred. As shown in Figure 10.1, energy recovery ranks below source reduction and recycling/reuse but above treatment and disposal. Incineration, which is confined and controlled burning, can not only decrease the volume of solid waste destined for landfills but can also recover energy from the waste burning process. This generates a renewable energy source and reduces carbon emissions by offsetting the need for energy from fossil sources and reduces methane generation from landfills (USEPA, 2020).

In a process known as "mass burn" MSW is unloaded from collection trucks at the MSW combustion facility and placed in a trash storage container. Then the trash

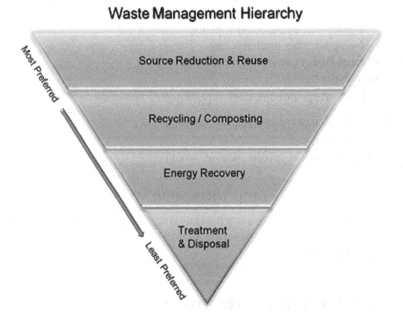

FIGURE 10.1   USEPA's non-hazardous waste management hierarchy. Source: USEPA (2020).

is sorted by an overhead crane, which then lifts it into a combustion chamber to be burned. The heat released from burning converts water to steam, which is then sent to a turbine generator to produce electricity. The remaining ash is collected and taken to a landfill where a high-efficiency baghouse filtering system captures particulates. As the gas stream travels through these filters, more than 99% of particulate matter is removed. Captured fly ash particles fall into hoppers (funnel-shaped receptacles) and are transported by an enclosed conveyor system to the ash discharger. They are then wetted to prevent dust and mixed with the bottom ash from the grate. The facility transports the ash residue to an enclosed building where it is loaded into covered, leak-proof trucks and taken to a landfill designed to protect against groundwater contamination. Ash residue from the furnace can be processed for the removal of recyclable scrap metals. Moreover, ash residue can be used for other beneficial reuse purposes discussed later.

In combusting MSW, there are three common technologies used: mass burning, modular systems, and refuse-derived fuel systems. *Mass burn facilities* are the most common type of combustion facility in the United States. These burn units burn MSW in a single combustion chamber under conditions of excess air. In combustion systems, excess air promotes mixing and turbulence to ensure that air can reach all parts of the waste. This is necessary because of the inconsistent nature of the waste. Most mass burn facilities burn MSW on a sloping, moving grate that vibrates or otherwise moves to agitate the waste and mix it with air. The waste used to fuel the mass burn facility may or may not be sorted before it enters the combustion chamber. Many advanced municipalities separate the waste on the front end to save recyclable products. The recyclable products/materials include:

- Newspapers
- Magazines
- Catalogs
- Junk mail
- Office paper
- Corrugated cardboard
- Gift and cereal boxes
- Phone books
- Paper bags
- Wrapping paper
- Aluminum bans
- Aluminum foil
- Plastic bottles
- Metal cans
- Metal pots

The non-recyclables that are not normally incinerated with MSW include:

- Ceramics
- Drinking glasses

- Bakeware
- Window glass
- Light bulbs
- Mirrors
- Plastic bags
- Styrofoam
- Paper towels
- Tyvek envelopes
- Wax covered boxes

In *modular systems* the burn process is different. The waste is burned unprocessed; it consists of mixed MSW. They also differ from mass burn facilities in that they are much smaller and portable. They can be moved from one site to another.

Refuse derived fuel systems use mechanical methods to shred incoming MSW, separate out non-combustible materials, and produce a combustible mixture that is suitable as a fuel in a dedicated incinerator or as a supplemental fuel in a conventional boiler system (USEPA, 2020).

BENEFICIAL REUSE OF MSW ASH

Incinerating MSW to use as an energy source is a beneficial reuse of a used, broken, and thrown-away material. Burning refuse significantly reduces the volume of the MSW and can result in gaining an energy source in the process. Sounds like a win–win situation whereby MSW is reduced substantially and is used as a feedstock for energy production. However, not only must costly air pollution control devices be part of the incineration process, but not all of the MSW that is burned is completely done away with, so to speak. After combustion of MSW, ash and slag remain and must be dealt with. The question becomes: Is there a beneficial reuse of ash available? The answer is yes.

An excellent example of using MSW ash for beneficial reuse has been practiced for more than 25 years by Hampton Roads Sanitation District (HRSD) in Hampton Roads Area (i.e., Norfolk–Virginia Beach–Chesapeake–Suffolk–Portsmouth region south of the Chesapeake Bay and Newport News–Hampton–Williamsburg north of the Chesapeake Bay), whereby the organization incinerates sewage sludge at three treatment plants in the area. Sewage sludge ash does not contain unburned metals like MSW ash does, so it is and can be used directly as an additive to concrete products such as revetments and novelty items.

Because of its proximity to the Chesapeake Bay, James River, York River, Elizabeth River, and the Atlantic Ocean where shore erosion is an ongoing problem, and based on the experience of other locations worldwide utilizing MSW and sludge-produced bottom ash beneficially, HRSD saw the benefit and practicality of doing likewise with its bottom ash produced in the Hampton Roads region.

HRSD's first beneficial reuse project dealt with providing bottom ash to companies that produce cement revetment structures for protection against wave-based erosion. Revetments are onshore structures with the principal function of protecting

FIGURE 10.2   Novelty turtle and frog made from bottom ash and cement.

the shoreline from erosion. Revetment structures, cast concrete slabs, typically consist of a cladding (covering) of asphalt, stone, and concrete to armor sloping natural shoreline profiles. Revetments are suitable for those sites with preexisting hardened shoreline structures. They work to mitigate wave action, require little maintenance, have an indefinite lifetime, and minimize impact to adjacent areas.

Another innovative use for bottom ash that HRSD has employed beneficially is in its creation of flowerpots and novelties such as reproductions of wildlife like turtles, rabbits, and other animals for display in home yards, gardens, or elsewhere (see Figures 10.1 and 10.2).

THE BOTTOM LINE

One might think that MSW facilities are common in the United States. Even though MSW facilities reduce the volume of waste and their byproducts, provide hot air for energy and ash for various applications, and are beneficial to the environment, this thought, or notion, is incorrect. So why are MSW incineration facilities not more common in the United States? Even though, in the United States, over 34 million tons of MSW were combusted in 2017 for energy use, MSW accounts for only a small portion of American waste management (USEPA, 2020). There are multiple reasons for this. One factor is a matter of space. For example, European countries are more likely to employ combustion of MSW because their space is limited and reduces the opportunities for other options. On the other hand, the United States encompasses a large amount of land and thus space limitations are not an important factor in the adoption of combustion with energy recovery. Landfilling in the United

States is usually considered a more viable option, especially in the short term, due to the low economic cost of building an MSW landfill versus the high cost of building incinerator systems.

Another factor in the slow growth of MSW combustion in the United States is public opposition to the facilities. Before air control technology, these facilities operated unabated with regard to polluting the local atmosphere. Thus, they gained a reputation as polluters. So, even today these facilities are frowned upon by many; they gained a bad reputation that is difficult to erase. Another local issue is traffic. Many communities do not want the increased truck traffic or to be adjacent to any facility handling municipal waste.

Moreover, another issue is cost. The upfront money needed to build an MSW combustion facility can be significant, and economic benefits may take several years to be fully realized. It typically requires an initial outlay of about 100 million dollars to finance the construction; however, larger plants may require double or triple that amount. To help recover costs, MSW combustion facilities typically collect a tipping fee from the independent contractors that drop the waste off on a routine basis. Income is also obtained from utilities after the electricity generated from the waste is sold to the grid. Another source of income is sometimes provided to MSW combustion facilities from the sale of both ferrous (iron) and non-ferrous scrap metals collected from the post-combustion as a stream.

The real bottom line is that whenever a waste product can be put to work as a beneficial resource or material, it is not "waste" but instead is feedstock for some other use.

REFERENCE

USEPA. (2020). Energy recovery from the combustion of municipal solid waste (MSW). Accessed 03/19/2021 @ https://www.epa.gov/smm/energy-recovery-combustion-muni cipal-solid-waste.msw.

11 Animal Waste

It's an industry, I no longer want to get tangled up in, even at the level of the ninety-nine-cent exchange. Each and every quarter pound of hamburger is handed across the counter after the following production costs, which I've searched out precisely: 100 gallons of water, 1.2 pounds of grain, a cup of gasoline, greenhouse-gas emissions equivalent to those produced by a six-mile drive in your average car, and the loss of 1.25 pounds of topsoil, every inch of which took 500 years for the microbes and earthworms to build. How can all this cost less than a dollar, and who is supposed to pay for the rest of it? If I were a cow, right here is where I'd go mad.

(Kingsolver, 2002, p. 120)

"Animals wastes are commonly considered the excreted material from live animals. However, under certain production conditions, the waste may also include straw, hay, wood shavings, or other sources of organic debris" (Encyclopedia.com, 2021).

FOOD RECOVERY HIERARCHY

From Figure 11.1, it is apparent and clear that feeding animals is the third tier of EPA's Food Recovery Hierarchy. Those involved with farming have been doing this for centuries. With proper and safe handling, anyone can donate food scraps to animals. Food scraps for animals can save farmers and companies' money. It is often cheaper to feed animals food scraps rather than having the scraps hauled to a landfill. Companies can also donate extra food to zoos or producers that make animal or pet food.

The point to be taken away from the Food Recovery Hierarchy is that all food waste is not always waste; it has beneficial reuse possibilities, such as feeding livestock. This is a good thing, of course. However, in this discussion, the focus is on waste produced by farm animals and its subsequent disposal. Note that the term "farm animals" is a broad, comprehensive, all-encompassing, and far-reaching term. For example, to make this point clear, consider the following list of "farm animals":

- Alpaca
- Banteng
- Beef cattle/cow
- Chicken
- Carp
- Camel
- Donkey
- Dog
- Duck
- Emu

DOI: 10.1201/9781003252665-11

- Goat
- Gayal
- Guinea pig
- Goose
- Horse
- Honeybee
- Llama
- Pig
- Pigeon
- Rhea
- Rabbit
- Sheep
- Silkworm
- Turkey
- Yak
- Zebu

One thing all the animals listed above have in common is their "natural" ability or tendency to produce waste of one form or another. The wastes or "organic by-products" produced naturally or contributed by animals include a variety of materials

FIGURE 11.1 EPA's Food Recovery Hierarchy. Source: USEPA, 2021.

such as solid and liquid animal manures, used bedding, spilled feed, and a variety of other substances. However, most livestock-associated organic by-products or waste are animal manures.

It would be an enormous undertaking to list all the natural and/or potential waste products that the list of animals above could contribute to the environment, in one fashion or the other; thus it is the focus of this chapter, instead, to narrow the discussion to a representative sample and to target beef cattle.

CATTLE WASTE

Although a high-risk enterprise, the United States is the leading beef producer in the world. Almost 30 billion pounds of beef were produced in the United States in 2019 and per capita consumption totaled 78 pounds. The beef industry is at high risk because the cattle cycle is risky, and, currently, the cycle is in a declining phase. During some years, an operation may not recover out-of-pocket costs. In the near term, several more years are expected of smaller calf crops, a slight decline in cattle feeding, a small decline in slaughter rates, and stable consumption rates. Profitability in the cattle business usually increases as production declines (PSU, 2005).

▶▶ **Interesting point**: Traditional feeder-cattle enterprises grow weaned calves (450–600 pounds) and yearling steers or heifers (550–800 pounds) to slaughter weights of 1,100–1,400 pounds.

This livestock sector discusses beef cattle feeding, confinement, and manure handling operations. This livestock sector includes adult beef cattle (heifers and steers) and calves. Beef cattle may be kept on open pastures or confined to feedlots. Note that herein only feedlot operations, excluding pasture operation, are discussed.

SIZE AND LOCATION OF INDUSTRY

There were more than 100,000 beef open feedlots in the United States, excluding farms where animals graze (USDA, 2020). (Note that 30,320 feedlots in the United States with less than 1,000 head of cattle compose the vast majority of U.S. feed lots (85.5%) (CME, 2016)). These feedlots sold more than 26 million beef cattle. Table 11.1 shows the distribution of feedlots by state and estimated capacity. The capacity of a beef feedlot is the maximum number of cattle that can be confined at any one time. The feedlot capacity was derived from annual sales figures by considering the typical number of turnovers of cattle per year and capacity utilization (ERG, 2020).

While most feedlots are small, the majority of production is from larger farms. For example, 2,075 feedlots with a capacity greater than 1,000 head accounted for only 2% of all lots but produced 80% of the beef sold in the United States in 2007. Beef feedlots vary in size from feedlots with a confinement capacity of less than 100 head to those in excess of 32,000 head of cattle.

TABLE 11.1

Number of Beef Feedlots by Size in 1997–2020

	Confinement Capacity		
	<500 Head	500–1,000 Head	>1,000 Head
Alabama	921	1	1
Alaska	19	0	0
Arizona	153	2	12
Arkansas	1,039	2	2
California	901	9	41
Colorado	1,400	44	145
Connecticut	151	0	0
Delaware	66	1	1
Florida	549	0	0
Georgia	696	1	2
Hawaii	34	1	3
Idaho	899	8	40
Illinois	7,184	54	51
Indiana	6,001	19	13
Iowa	12,040	233	263
Kansas	2,630	93	298
Kentucky	1,910	6	4
Louisiana	311	0	0
Maine	243	0	0
Maryland	754	1	0
Massachusetts	111	0	0
Michigan	4,455	21	30
Minnesota	8,345	58	56
Mississippi	560	0	0
Missouri	4,392	16	23
Montana	655	14	16
Nebraska	4,855	204	602
Nevada	83	4	4
New Hampshire	79	0	0
New Jersey	335	0	0
New Mexico	321	3	16
New York	1,424	2	3
North Carolina	903	2	3
North Dakota	1,086	9	8
Ohio	7,241	19	11
Oklahoma	1,850	11	35
Oregon	1,864	5	11
Pennsylvania	5,299	16	10
Rhode Island	26	0	0

(Continued)

TABLE 11.1 (CONTINUED)
Number of Beef Feedlots by Size in 1997–2020

State	Confinement Capacity		
	<500 Head	500–1,000 Head	>1,000 Head
South Carolina	348	3	1
South Dakota	2,711	65	88
Tennessee	1,965	1	1
Texas	3,574	31	218
Utah	797	5	11
Vermont	158	1	1
Virginia	1,363	4	3
Washington	1,170	4	22
West Virginia	804	0	0
Wisconsin	7,980	19	10
Wyoming	345	8	16
United States	**103,000**	**1,000**	**2,075 ‖ 106,075**

OERG, 2020

Beef cattle are located in all 50 of the United States, but most of the capacity is in the central and western states.

BEEF CATTLE SECTOR PROFILE

Cattle production accounted for $66.2 billion in cash receipts in 2019, making it the most important agricultural industry in the United States. Of the total cash receipts forecast for agricultural commodities in 2019, cattle production represented about 18% of the $375 billion in total. With its rich agricultural land resources, the United States has developed a beef industry that is largely separate from its dairy sector. In addition to having the world's largest fed-cattle industry, the United States is also the world's largest consumer of beef—primarily high-value, grain-fed beef. The beef cattle sector is roughly divided into two production sectors: cow-calf producers and cattle feeding.

BEEF PRODUCTION OPERATIONS

There are three different types of operations in the beef industry with each corresponding to a different phase of the animal growth cycle. These operations are referred to as cow-calf operations, backgrounding, and finishing. These operations are typically conducted at separate locations that specialize in each phase of production.

Cow-Calf Operations

Beef cow-calf production is relatively widespread and economically important in most of the United States. According to the USDA 2004 *Census of Agriculture*, about a million farms had inventories of cattle and calves that generated more than $41 billion in sales, accounted for more than 21% of the total market value of agricultural products sold in the United States, and ranked first in sales among all commodities.

Cow-calf types of operations are a source of the heifers and steers (castrated males) fed for slaughter. Cow-calf operations maintain a herd of heifers, brood cows, and breeding bulls typically on pasture or range land, to produce a yearly crop of calves for eventual sale as feeder cattle. In colder climates and during drought conditions, cow-calf operations using pasture or rangeland will provide supplemental feed, primarily hay but with some grains and other feedstuffs. Confinement on dry lots also is an option used on some cow-calf operations when grazing will not satisfy nutritional needs. Although pasture or range-based cow-calf operations are most common, operations exclusively using dry lots may be encountered. In colder climates, cow-calf operations may have calving barns to reduce calf mortality.

Background Operations

Backgrounding or stocker operations describes a management system where recently weaned calves or yearling cattle are grazed for finishing on high-energy rations to promote rapid weight gain for a period of time before they are placed in the feed yard. Backgrounding operations may be pasture or dry-lot based or some combination thereof. Relatively inexpensive forages, crop residues, and pasture are used as feeds with the objective of building muscle and bone mass without excessive fat at a relatively low cost. The length of the backgrounding process may be as short as 30–60 days or as long as 6 months (Ransby et al., 1996). The duration of the backgrounding process and the size of the animal moving onto the finishing stage of the beef production cycle depend on several factors. High grain prices favor longer periods of backgrounding by reducing feed costs for finishing or fattening while heavier weaning weights shorten the finishing process. Backgrounded beef cattle may be sold to a finishing operation as "feeder cattle," usually at an auction or raised under contract with a finishing operation. It is common for large finishing operations to have cattle backgrounded under contract to ensure a steady supply of animals. In some instances, cow-calf and backgrounding operations will be combined.

Finishing or Feedlot Operations

The final phase of the beef cattle production cycle is called the finishing or feedlot phase. Beef cattle in the finishing phase are known as "cattle on feed." Finished cattle are "fed cattle." Usually, the finishing phase begins with six-month-old animals weighing about 400 pounds. In between 150 and 180 days, these animals will reach the slaughter weights of 1,050–1,150 pounds for heifers and 1,150 and 1,250 pounds for steers and a new finishing cycle begins. Some feedlot operators will start with younger animals weighing about 275 pounds, or older or heavier animals initially. This either extends the finishing cycle to about 270 days or shortens it to about

100 days. Accordingly, typical feedlots can have from 1.5 to 3.5 turnovers of cattle herds. On average, most beef feedlots operate at between 80% and 85% of capacity over the course of a year (USDA, 2021).

BEEF CONFINEMENT PRACTICES

The cow-calf and backgrounding phases of the beef production cycle are primarily pasture or rangeland based. The underlying rationale for this method of raising cattle is avoidance of the cost of harvesting, transporting, and storing roughages, which are necessary with confinement feeding. Therefore, confinement feeding during these phases of the beef production cycle generally is limited to time periods when grazing cannot satisfy nutritional needs.

In the final or finishing phase of the beef cattle production cycle, heifers and steers most typically are fed to slaughter weight in open confinement facilities known as feedlots or feed yards. The majority of beef feedlots are open feedlots, which may be partially paved. Generally, paving, if present, is limited to a concrete apron typically located along feed bunks and around waterers, because these are areas of heaviest animal traffic and manure accumulation (Bodman, et al., 1987).

Cattle are segregated in pens designed for efficient movement of cattle, optimum drainage, and easy feed truck access. A typical pen holds 150–300 head of cattle, but the size can vary substantially. Required pen space may range from 75 to 400 sq.ft of pen space per head, depending on the climate. A dry climate requires 75 sq.ft of pen space per head whereas a wet climate may require up to 400 sq.ft (Thompson, O'Mary, 1983). Space needs vary with the amount of paved space, soil type, drainage, annual rainfall, and freezing and thawing cycles. These types of operations may use mounds to improve drainage and provide areas that dry quickly, since dry resting areas improve cattle comfort, health, and feed utilization. Typically, pens are constructed to drain as quickly as possible after precipitation events with the resulting runoff conveyed to storage ponds that may be preceded by settling basins to reduce solids entering the ponds. In open feedlots, protection from the weather is often limited to a windbreaker near a fence in the winter and/or sunshade in the summer.

In cold climates and high rainfall areas, small beef cattle finishing operations may use totally enclosed confinement to reduce the negative impact of cold weather on feed conversion efficiency and rate of weight gain. However, totally enclosed confinement facilities generally are not economically competitive with open feedlots and are relatively few in number.

FEEDING PRACTICES

Feeding practices in the different phases of the beef production cycle differ, reflecting differences in nutritional requirements for maintenance and growth. As mentioned, cow-calf and backgrounding operations typically depend on grazing, possibly with the feeding of a mineral supplement to satisfy nutritional needs. When there is feeding in confinement facilities, harvested roughages, hays, and silages are the principal, if not only, feedstuffs.

During the finishing phase of the beef production cycle, there is a shift from a roughage-based to a grain-based, high-energy ration to produce a rapid weight gain and desirable carcass characteristics. Because beef cattle are ruminant animals, some small level of roughage intake must be maintained to maintain rumen activity. Generally, mixed rations, which are combinations of roughages and concentrates, are fed. However, roughages and concentrates may be fed separately, a practice more common with smaller operations. Roughages have high fiber contents and are relatively dilute sources of energy and protein, whereas concentrates are low-fiber, high-energy feeds, which also may have a high protein content. Feeding practices for beef cattle generally are based on nutrient requirements established by the National Research Council (NRC, 1996).

> ▶ **Key term**: The *rumen* is a large, hollow muscular organ. It is one of the four stomach compartments in ruminant animals. A fermentation vat, the rumen can hold 160–240 liters of material and is the site of microbial activity.
> ▶▶ **Important point**: Handling moist feeds have a limited potential for particulate emissions, while handling dry feeds, such as grain, may be a source of particulate emissions.

While cow-calf and backgrounding operations generally depend on grazing to satisfy nutritional needs, feed must be provided to beef cattle being finished in feedlots. Typically, feed is delivered to feed bunks two to three times per day with the objective of always having feed available for consumption without the excessive accumulation of uneaten feed to minimize spoilage. Cattle are typically fed using feed bunks located along feed alleys that separate individual pens. Feed is delivered either by self-unloading trucks or tractor-drawn wagons (fence-line feeding) or mechanical feed bunks. Usually, mechanical feed bunks are located between pens, allowing animal access from both sides of the feed bunk. In small feedlots where roughages and concentrates are fed separately, animals may have access to haystacks, self-feeding horizontal silos, or large tubular plastic bags containing roughage. Concentrates are fed separately in portable feed tanks.

Open-front barns and lots with mechanical or fence-line feed bunks are common for feedlots up to 1,000 head, especially in areas with severe winter weather and high rainfall. Portable silage and grain bunks are useful for up to 200 heads (Bodman, et al., 1987).

The metabolic requirements for maintenance of an animal typically increase during cold weather, reducing weight gain and increasing feed consumption to provide more energy, thereby increasing the amount of manure that is generated. Feed consumption typically declines under abnormally high temperatures, therefore reducing weight gain. Investigations in California have shown that the effect of climate-related stress could increase feed requirements as much as 33%, resulting in increased manure generation (Thompson, O'Mary, 1983).

MANURE MANAGEMENT PRACTICES

Beef cattle manure produced in confinement facilities generally is handled as a solid. Runoff from feedlots can be either liquid or slurry. Manure produced in totally enclosed confinement facilities may be handled as slurry or a liquid if water is used to move manure. Slurry manure has enough water added to form a mixture capable of being handled by solids handling pumps. Liquid manure usually has less than 8% solids resulting from significant dilution. It is easier to automate slurry and liquid manure handling, but the large volume of water necessary for dilution increases storage and disposal requirements and equipment costs (USDA, 1995).

Solid manure is scraped or moved by tractors to stockpiles. Runoff from open lots is pumped to solids separation activities to separate the solid and liquid fraction. The liquid fraction is then sent to storage ponds. Both the solid and liquid fractions can be disposed of on land.

MANURE COLLECTION

The following methods are used in feedlots to collect accumulated manure for disposal:

Open lots: Manure most commonly is collected for removal from open lots by scraping, using tractor-mounted blades. Very large feedlots commonly use earth-moving equipment such as pan scrapers and front-end loaders. Manure accumulates in areas around feed bunks and water troughs most rapidly, and these areas may be scraped frequently during the finishing cycle. This manure may be removed from the pen immediately or may be moved to another area of the pen and allowed to dry. Usually, the entire pen is completely scraped, and the manure is removed at the end of finishing after the animals are shipped for slaughter (Sweeten, 2000).

Totally enclosed confinement: Beef cattle manure accumulations in totally enclosed confinement facilities also are typically collected and removed by scraping, using tractor-industry technology. Scrapers also can be used but require a concrete floor. With a concrete floor, the use of a flush system for manure collection and removal also is possible. A flush system uses a large volume of water discharged rapidly one or more times per day to transport accumulated manure to an earthen anaerobic lagoon for stabilization and storage. Typically, 100 gallons of flush water is used per head twice a day. The frequency of flushing, as well as slope and length of the area being flushed, determines the amount of flush water required (Loudon et al., 1985). The lagoon usually is the source of the water used for flushing. Due to freezing problems, the use of flushing in totally enclosed finishing facilities is not common since totally enclosed confinement operations normally are found only in cold climates.

Slatted floors over deep pits or shallow, flushed alleys also have been used in totally enclosed beef cattle finishing facilities. Most slats are reinforced concrete but can also be wood, plastic, or aluminum. They are designed to support the weight of the slat plus a live load, which includes animals, humans, and mobile equipment. Manure is forced between the slats as the animals walk around the facility, which keeps the floor surface relatively free of accumulated manure. With slatted floors over deep pits, pits typically are emptied at the end of a finishing cycle. Some water may be added to enable pumping, or the area may be accessed to allow the use of a front-end loader. Due to the cost of slatted floor systems, their use in beef cattle production is rare.

Factors that affect emissions from beef feedlots include the number of animals on the lot and the moisture of the manure. The number of animals influences the amount of manure generated and the amount of dust generated. In well-drained feedlots, emissions of nitrogen oxides are likely to occur because decomposition of manure is aerobic. In wet feedlots, decomposition is anaerobic and emissions of ammonia, hydrogen sulfide, and other odor-causing compounds are likely. Additionally, the feedlot is a potential air release point of particulate matter/dust from feed and movement of cattle.

Manure Storage, Stabilization, Disposal, and Separation

Manure collected from the feedlot may be stored, stabilized, directly applied to land on-site, or transported off-site for disposal.

Storage

If beef cattle manure is handled as a solid, it is stored by stacking within an area of the feedlot or other open confinement facility or an adjacent dedicated storage site. Stacking sites typically will be uncovered, and collection of contaminated runoff is necessary. Manure handled as a slurry or liquid will be stored in either earthen storage ponds or anaerobic lagoons. Above-ground tanks are another option for storage of these types of manures but are not commonly used. Storage tanks and ponds are designed to hold the volume of manure and process wastewater generated during the storage period, at the depth of normal precipitation minus evaporation, and the depth of the 25-year, 24-hour storm event with a minimum of 1 ft of freeboard remaining at all times. Emissions from storage tanks and ponds include ammonia, hydrogen sulfide, VOC, and methane. The magnitude of emissions depends primarily on the length of the storage period and the temperature of the manure. Low temperatures inhibit the microbial activity responsible for the creation of these compounds, while long storage periods increase the opportunity for emissions.

Stabilization

Stabilization is the treatment of manure to reduce odor and volatile solids prior to land application. Because manure is allowed to remain on feedlots for extended time periods, a significant degree of decomposition due to microbial activity occurs. When stacked for storage, a significant increase in temperature may occur depending on moisture content due to microbial heat production. Manure accumulations on feedlots and stored-in stacks can be sources of ammonia, hydrogen sulfide, VOC, and methane if the moisture content is sufficient to promote microbial decomposition. Dry manure is an emission source of nitrous oxide and particulate matter/dust emissions. When beef cattle manure is stored as a slurry or liquid, some decomposition or stabilization also occurs. Anaerobic lagoons, when designed and operated properly, result in a higher degree of stabilization than storage ponds or tanks, which have the single objective of providing storage. In storage ponds and tanks, intermediates in the decomposition process usually accumulate and are sources of odors. Storage tanks and ponds and lagoons can be sources of ammonia, hydrogen sulfide, VOC, and methane emissions.

Land Application

The majority (approximately 83%) of beef feedlots dispose of their manure by storage and stabilization through land application (USDA, 2020). Box-type manure spreaders are used to apply solid manure while flail-type spreaders or tank wagons with or without injectors are used with slurry-type manure. Tank wagons or irrigation systems are used for liquid manure disposal. Beef cattle manure not disposed of by land application may be composted for sale for horticultural and landscaping purposes.

Separation

In the beef cattle industry, liquid–solids separation essentially is limited to the removal of solids from runoff collected from feedlots and other open confinement areas using settling basins. However, stationary and mechanical screens also may be used. The objective of these devices is to reduce the organic loading to runoff storage ponds. Although separation also can be used with beef cattle manure handled as a liquid, this form of manure handling is not common in the beef cattle industry, as noted earlier. Emissions from settling basins depend on the hydraulic retention time (HRT) of the runoff in the basin and the frequency of removal of settled solids. If settled solids are allowed to accumulate, ammonia, hydrogen sulfide, VOC, and methane emissions may be significant. Generally, the time spent in separation activities is short (i.e., less than one day).

BEEF VIRTUAL FARMS

Virtual farms are hypothetical farms that are intended to represent the range of design and operating practices that influence emissions from each animal sector. These virtual models can be used to develop mission estimates, control costs, and regulatory assessments.

The virtual farms include four components: confinement areas, solids separation activities, storage and stabilization practices, and land application. Land application includes emissions from the manure application activity and from the soil after manure application. For the virtual farms, emissions from the application of manure are differentiated from emissions from the manure application site (i.e., cropland or other agricultural lands) because emission mechanisms are different. Emissions from the application activity occur in a short time period and depend on the methods by which manure is applied. Emissions from the application site occur as substances volatize from the soil over a period of time as a result of a variety of subsequent microbial and chemical transformations.

Cow-calf and background operations do not typically confine animals and, as such, virtual models were not developed to represent them. Those that do confine cattle would be represented by the virtual model farms for finishing operations.

Two virtual farms were developed to characterize typical beef cattle finishing operations. The components of the virtual farms include an open confinement area (feedlot), solids separation for collected surface runoff, manure storage facilities (storage ponds for surface runoff and stockpiles for solids), and land application. In both virtual models, land application includes solid and liquid manure application activities (e.g., irrigations and solid manure spreader) and the manure application site (e.g., emission released from agricultural soils after the manure is applied). The beef virtual models differ only by the presence or absence of solids separation.

CONFINEMENT

Feedlots are the only confinement operation considered for the virtual model farms because most, if not at all, beef operations use feedlots. Industry manure collection information indicates that most of the manure is typically scraped by a tractor scraper or front-end loader and stockpiled for later disposal by land application. Runoff from the feedlot is sent to solids separation processes or directly to storage ponds.

SOLIDS SEPARATION

Runoff from the feedlot is either sent to solids separation activities to remove solids or directly to storage ponds. The separated solids are sent to a stockpile and the liquid fraction is sent to a storage pond. Two common types of solids separation were considered in developing the virtual model farms: mechanical screens or gravity settling basins. After reviewing the emission mechanisms from each type of separation practice, it was determined that emission should not vary substantially between mechanical screens and settling basins. Additionally, due to the short duration, manure emissions would be relatively small, thus differences between the separation processes would be insignificant. Therefore, the model virtual farms only represent the option of either having solids separation or not. The virtual models are based on a short manure retention time in solids separation, and therefore negligible emissions from this process. The emission differences between the models are from the manure storage following separation.

STORAGE AND STABILIZATION

The virtual model farms contain storage activities for solid and liquid manure. Two types of solid manure storage activities were considered in developing the virtual model farms. Solid manure could be: (1) stored in an uncovered stockpile, or (2) not stored at all and sent directly from the feedlot to be land applied. The review of industry practices indicated that solid manure would generally not be sent directly from the feedlot to be land applied but would have some intermediate storage. Therefore, all the model farms included an uncovered stockpile. The liquid fraction from the runoff or the solids separation process is sent to a storage pond.

LAND APPLICATION

As previously mentioned, land application includes the manure application activity and the manure application site (i.e., cropland or other agricultural lands). Solid manure is typically land applied to the manure application site using a solid manure spreader. Three types of land application activities were considered for liquid manure in developing the virtual model farms: land application by (1) liquid surface spreader, (2) liquid injection manure spreader, or (3) irrigation. The review of industry practices indicated that injection is rarely used. The emissions from irrigation and liquid surface spreading were judged to be similar, due to the short duration of time for each activity and similar emission mechanisms. Therefore, the virtual model farms only refer to liquid manure land application rather than a specific type.

WHEN ANIMAL WASTE IS NOT WASTED

Beef manure (and other animal manure) is viewed as a waste. In the past, farmers have viewed animal manure as a valuable resource. However, with the advent of increasing concerns about the environment, increased passage of environmental regulations, and hardened views on animal manure as nothing more than a headache to deal with or dispose of, the view on animal waste changed from that of a beneficial resource to nothing more than a waste. The truth be told, the tendency to designate livestock manure as a "waste" has led to the undervaluation of manure as a source of nutrients. Simply, classifying livestock manure as a waste implies the substance has no value. This implication is incorrect. It is more correct to replace the word "waste" with the terms manure, residuals, or by-products. Farmers and agricultural experts have for years known the value of livestock manure.

So what is the value of livestock manure, residual, or by-product? When properly used, animal manure is a resource and should be regulated as such. Many benefits are to be gained by properly applying livestock manure to land. These benefits include:

- Serves as a fertilizer
- Increases crop quality
- Maintains soil pH
- Increases soil organic matter
- Increases soil physical properties
- Decreases pesticide dependence
- Reduces runoff and soil loss
- Sequesters carbon

FERTILIZER

Livestock manure is an excellent source of major plant nutrients such as nitrogen (N), phosphorus (P), and potassium (K); it also provides many secondary nutrients that plants require. The actual nutrient value of livestock manure differs considerably due to the type of animal. For example, the nutrient analysis for beef in nonliquid systems (kg/Mg) is 2–10 nitrogen versus a nitrogen level for poultry of 2–66 in a nonliquid system Bates and Gagon, 1981). In addition, the actual nutrient value of livestock manure differs due to livestock food ratio, method of collection and storage, and method application and climate. Note that nutrients in manure may be lost or transformed during treatment, storage, and handling, affecting their availability for use by growing plants.

CROP QUALITY

Because livestock manure contains most of a plant's essential elements, it is an excellent nutrient source (Follett et al., 1992). Even though the concentrations of nutrients in manure tend to be low, it is a great source of plant nutrients for crop production and crop quality (Eck et al., 1990). Many studies have shown that land application of manure produces crop yields equivalent or superior to those obtained with chemical fertilizers (Motavalli et al., 1989). When crop improvements with manure were greater than those attained with commercial fertilizer, the response was usually attributed to manure-supplied nutrients or to improved soil conditions not provided by commercial fertilizer (CAST, 1992).

SOIL pH

Livestock manure application to soil raises soil pH. The main reason manure raises soil pH is due to calcium (Ca) and magnesium (Mg) contained in the manure. Thus, applying livestock manure to acid soils not only supplies much-needed nutrients and organic matter for plant growth but also reduces soil acidity, thus improving available P and reducing aluminum (Al) toxicity.

SOIL ORGANIC MATTER

Research has shown that an increase in soil organic matter (SOM) with manure applications has a significant effect on the chemical, physical, and biological properties of the soil (Haynes and Naidu, 1998). The organic matter deposited enhances soil physical properties such as tilth (tilled soil), structure, water-holding capacity, water infiltration rate, and soil microbial activity (Sweeten and Mathers, 1985). Studies have shown that the increase in SOM are directly related to manure addition (Aoyama et al., 1999; Brown et al., 2000; Fraser et al., 1988). The greater plant growth is in response to nutrients from manure which work to increase root biomass and residues in SOM.

PHYSICAL SOIL PROPERTIES

Manure has the ability to promote the formation of water-stable aggregates (WSA) in soil. This capability has a profound effect on soil structure and thus on soil physical characteristics (Hanes and Naidu, 1998). A high percentage of WSA increases infiltration (Roberts and Clanton, 2000), porosity (Kirchmann and Gerzabek, 1999), and water holding capacity (Mosaddeghi et al., 2000). WSA is also associated with decreased compaction and erosion (Barthes et al., 1999; Mosaddeghi et al., 2000). Note that Angers (1998) reported that even in silty clay soils with high organic matter contents, the addition of manure increases macro-aggregation, which helps prevent structural degradation.

PESTICIDE DEPENDENCE

The soil food web is affected by the presence and forms of organic carbon in the soil. This in turn affects the structure and complexity of the soil food web, affecting nutrient cycling and both plant diseases and parasites. Moreover, research indicates that microbial biomass and activity increase with manure additions (Estevez et al., 1996; Haynes and Naidu, 1998; Kandeler et al., 1999). Shifts in nutrient cycling occur with increases in microbial populations. Experience has shown that manure additions increase bacteria involved in the nitrogen cycle (see Figure 11.2). High nitrogen manure can suppress diseases by generating high ammonia and/or nitrous acid concentrations in the soil (Lazarovits, 2001). Organic matter is known to affect activity, degradation, and persistence of pesticides. Land application of composted rather than fresh manure has the potential for reducing weed seed viability and thus reducing herbicides needed (Edwards et al., 1994). The advantage the compost offers lies in the fact that the variety of microbial and biological stimulants in the compost may actually inoculate the soils.

REDUCING RUNOFF AND SOIL LOSS

In addition to improving infiltration, aggregation, and bulk density of soil, manure application can also affect soil erodibility (Mielke and Mazurak, 1976; Sommerfeldt et al., 1988). These are changes in soil properties that can have a substantial impact on the runoff and soil loss from fields where manure has been land applied. These impacts have been verified and substantiated based on study and experience. Runoff and erosion rates were found to be influenced by manure characteristics, loading rates, incorporation, and the time between application and the first rainfall.

SEQUESTERING CARBON

Land application of livestock manures could help mitigate potentially negative consequences of rising atmospheric carbon dioxide on the global climate by contributing to greater sequestration of carbon in soil (CAST, 1992). In general, soil organic carbon sequestration on a land area basis appears to be greater with an increased rate of manure application (Sommerfeldt et al., 1988; Gupta et al., 1992).

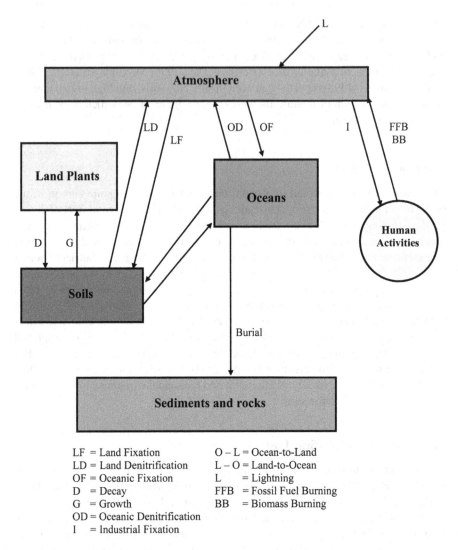

FIGURE 11.2   Nitrogen cycle box model.

LF = Land Fixation   O – L = Ocean-to-Land
LD = Land Denitrification   L – O = Land-to-Ocean
OF = Oceanic Fixation   L   = Lightning
D = Decay   FFB   = Fossil Fuel Burning
G = Growth   BB   = Biomass Burning
OD = Oceanic Denitrification
I = Industrial Fixation

DID YOU KNOW?

Livestock manure benefits soils and plants beyond its ability to supply nitrogen, phosphorus, potassium, and micronutrients. Soil's porosity, structure, water infiltration rate, and moisture retention capacity are all improved with manure application. The positive benefits of manure application to soils may take two or more years before improvement is evident (Sweeten and Mathers, 1985).

THE BOTTOM LINE

When entering a fenced cow pasture to hike across it or around it or to simply enter and take in whatever view is obtainable, it is common practice for each of us to pay attention to what is lying on the ground. Why? Well, if there are cattle or livestock of any kind corralled inside the pasture, they are going to consume whatever is available to eat and drink. When Nature calls, they are going to defecate and maybe perform this function several times each day. Because of their need to live up to the biological fact that what goes in must come out (to a point, that is), a liquid deposit of manure is deposited, forming what is commonly called a cow pie. Again, while in the pasture with a few cows, it is wise to look where one walks. When a cow pie is spotted, it is quickly avoided so as not to step in the mess. Actually, those who avoid such messes realize they are simply stepping away from animal waste.

But is livestock dung, droppings, or cow pie residue really waste? During the Western era in the United States, it was not uncommon for pioneers, Native Americans, and settlers to collect buffalo and other animals' dung to provide fuel for the fires needed to cook and keep warm. After burning the dung, the ashes left over simply were absorbed into the soil of the fire pits or were borne on the wind and transplanted wherever it landed, acting as a soil amendment and providing carbon to the soil.

So the point is this, animal waste used in agriculture or to provide a fuel source for those who need to cook and keep warm in the wild is a waste that is not wasted; instead, it serves a beneficial reuse function.

REFERENCES AND RECOMMENDED READING

Aoyama, M., Angers, D.A., N'Dayegamiye, A., and Bissonnette, N. (1999). Protected organic matter in water-stable aggregates as affected by mineral fertilizer and manure applications. *Canadian Journal of Soil Science*, 79(3): 419–425.

Barthes, B., Albrecht, A., Asseline, J., De Noni, G., and Roose, E. (1999). Relationship between soil erodibility and topsoil aggregate stability or carbon content in a cultivated Mediterranean highly (Aveyron France). *Communications in Soil Science and Plant Analysis*, 30(13–14): 1929–1938.

Bates, T., and Gagon, E. (1981). *Nutrient Content of Manure*. Ontario, Canada: University of Guelph.

Bodman, G.R., Johnson, D.W., Jedele, D.G., Meyer, B.M., Murphy, J.P., and Person, H.L. (1987). Manure accumulation. Accessed 02/02/2021 @ https://www.valleyair.org/notices/docs/2020.

Brown, S.M.A., Cook, H.F., and Lee, H.C. (2000). Topsoil characteristics from a paired farm survey of organic versus conventional farming in southern England. *Biological Agriculture & Horticulture*, 18(1): 37–54.

CAST. (1992). Preparing U.S. agriculture for global climate change. *Task Force Report No.* 119, Council for Agricultural Science and Technology, Ames, IA. 96 pp.

CME, Chicago Mercantile Exchange (2016). *Cattle market volatility*. Accessed 02/02/2021 @ https://www.theobia.com/files.

Eck, H.V., Winter, S.R., and Smith, S.J. (1990). Sugarbeet yield and quality in relation to residual beef feedlot waste. *Agronomy Journal*, 82: 250–254.

Edwards, D.R., Coyne, M.S., Daniel, T.C., Vendrell, P.F., Murdoch, J.F., and Moore, Jr., P.A. (1994). Indicator bacteria concentration of two northwest Arkansas streams in relation to flow and seas. *Transactions of the American Society of Agricultural Engineers*, 40: 103–109.

Encycopedia.com. (2021). Animal waste. Accessed 03/20/2021 @ https://www.encyclopedia.c om/environment/encyclopedia-almanacs-transcripts-and-maps/animal-waste.

ERG. (2020). *Facility Counts for Beer, Dairy, Veal, and Heifer Operations*. Memorandum form Deb Bartram, Eastern Research Group, Inc. (ERG) to the Feedlots Rulemaking Record. U.S. Environmental Protection Agency (USEPA) Water Docket, W-00–27.

Estevez, B., N'Dayegamiye, A., and Coderre, D. (1996). The effect of earthworm abundance and selected soil properties after 14 years of solid cattle manure and NPKMg fertilizer application. *Canadian Journal of Soil Science*, 76(3): 351–355.

Follett, R.H., Westfall, D.G., and Croissant, R.L.. (1992). Utilization of animal manure as fertilizer. Colorado State University Cooperative Extension Bulletin 552A, Colorado Springs, CO: Colorado State University.

Fraser, D.G., Doran, J.W., Sahs, W.W., and Lesoing, G.W. (1988). Soil microbial populations and activities under conventional and organic management. *Journal of Environmental Quality*, 17(4): 585–590.

Gupta, A.P., Narwal, R.P., Antil, R.S., and Dev, S. (1992). Sustaining soil fertility with organic-C, N.P, and K by using farmyard manure and fertilizer-N in a semiarid zone: A long-term study. *Arid Soil Research and Rehabilitation*, 6: 243–251.

Haynes, R.J., and Naidu, R. (1998). Influence of lime, fertilizer and manure applications on soil organic matter content and soil physical conditions: A review. *Nutrient Cycling in Agroecosystems*, 51: 123–137.

Kandeler, E., Stemmer, M., and Kilmanek, E.M. (1999). Response of soil microbial biomass, urease and xylanse within particle size fractions to long-term soil management. *Soil Biology and Biochemistry*, 32(2): 261–273.

Kingsolver, B. (2002). *Small Wonder*. New York: Perennial: HarperCollins.

Kirchmann, H., and Gerzabek, M.H. (1999). Relationship between soil organic matter and micropores in a long-term experiment at Ultuna, Sweden. *Journal of Plant Nutrition and Soil Science*, 162(5): 493–498.

Lazarovits, G. (2001). Management of soil-borne pathogens with organic soil amendments: A disease control strategy salvaged from the past. *Canadian Journal of Plant Pathology*, 23: 1–7.

Loudon, T.L., Jones, D.D., Petersen, J.B., Backer, L.F., Brugger, M.F., Converse, J.C., Fulhage, C.D., Lindley, J.A., Nelvin, S.W., Person, H.L., Schulte, D.D., and White, R. (1985). *Livestock Waste Facilities Handbook*, 2nd ed. Ames, IA: Midwest Plan Service.

Mielke, L.N., and Mazurak, A.P. (1976). Infiltration of water on a cattle feedlot. *Transactions of ASAE*, 19(2): 341–344.

Mosaddeghi, M.R., Hajabbasi, M.A., Hemmat, A., and Afyuni, M. (2000). Soil compactibilty as affected by soil moisture content and farmyard manure in central Iran. *Soil & Tillage Research*, 55(1–2): 87–97.

Motavalli, P.P., Kelling, K.A., and Converse, J.C. (1989). First-year nutrient availability from injected dairy manure. *Journal of Environmental Quality*, 18: 180–185.

NCBA. (1999). *Comments on the Draft Industry Profile*. National Cattlemen's Beef Association (NCBA).

NRC. (1996). *Nutrient Requirements of Beef Cattle*. National Research Council (NRC). Subcommittee on Beef Cattle Nutrition, Committee on Animal Nutrition, Board on Agriculture, 7th rev. ed.

PSU. (2005). *Feeding Beef Cattle*. Penn State University. Accessed 5/5/05 @ http://agalterna-tives-aers.psu.edu.

Rasby, R., Rush, I., and Stock R. (1996). *Wintering and Backgrounding Beef Calves.* NebGuide. Lincoln, NE: Cooperative Extension, Institute of Agriculture and Natural Resources, University of Nebraska-Lincoln.

Roberts, R.J., and Clanton, C.J. (2000). Surface seal hydraulic conductivity as affected by livestock manure application. *Transactions of the American Society of Agricultural Engineers*, 43(3): 603–613.

Sommerfeldt, T.G., Change, C., and Entz, T. (1988). Long-term annual manure applications increase soil organic matter and nitrogen and decrease carbon to nitrogen ratio. *Soil Science Society of America Journal*, 52: 1668–1672.

Sweeten, J. (2000). *Manure Management for Cattle Feedlots.* Great Plains Beef Cattle Handbook. Lincoln, NE: Cooperative Extension service – Great Plains States.

Sweeten, J.M., and Mathers, A.C. (1985). Improving soils with livestock manure. *Journal of Soil and Water Conservation*, 40(2): 206–210.

Thompson, G.B., and O'Mary, C.C. (1983). *The Feedlot*, 3rd ed. Philadelphia, PA: Lea & Febiger.

USDA. (1995). *Agricultural Waste Management Field Handbook, National Engineering Handbook, Part 651.* Washington, DC: U.S. Department of Agriculture (USDA), Natural Resources Conservation Service (NRCS).

USDA. (1999a). *Cattle: Final Estimates 1994–1998.* Statistical Bulletin 953. Washington, DC: U.S. Department of Agriculture (USDA), National Agricultural Statistics Service (NASS).

USDA. (1999b). *1997 Census of Agriculture.* Washington, DC: U.S. Department of Agriculture (USDA), National Agricultural Statistics Service (NASS).

USDA. (2020). *Agricultural Waste Management Field Handbook, National Engineering Handbook, Part 651.* Washington, DC: U.S. Department of Agriculture (USDA), Natural Resources Conservation Service (NRCS).

USEPA. (2001a). *Emissions from Animal Feeding Operations.* Research Triangle Park, NC: United States Environmental Protection Agency, Office of Air Quality Planning and Standards.

USEPA. (2001b). *Development Document for the Proposed Revisions to the National Pollutant Discharge Elimination System Regulation and the Effluent Guidelines for Concentrated Animal Feeding Operations.* EPA-821-R-01-003. Washington, DC: U.S. Environmental Protection Agency, Office of Water.

USDA. (2021). *National Animal Health Monitoring System, Part 1: Baseline Reference of Feedlot Management Practices.* Fort Collins, CO: U.S. Department of Agriculture (USDA), Animal and Plant Health Inspection Service (APHIS).

12 Human Waste
The Waste Cloud

It is common practice when treating wastewater to churn out effluent water cleaner than the local waterways it is ultimately outfalled into. It is ultimately sent on to the ocean with no downstream use—in other words "one and done" usage. Why? Why waste such a valuable resource? Why not reuse it? Don't we already reuse it … in de facto water recycling?

F. R. Spellman (2017)

NIGHT SOIL

The caption for Figure 12.1 points out what the author calls a cloud of wastes by other names. Note that the entries included in the "cloud" are not all inclusive—an entire listing of the different types of waste is much larger.

Anyway, this chapter deals with human waste. Okay, let's be clear we are not discussing substances, materials, objects, or things that humans waste (use, destroy, or throw away). No. This discussion is about metabolic waste in the form of excretion but not carbon dioxide from the lungs but instead the nitrogenous waste known as urine. Moreover, that other familiar and detested human waste product has to do with defecation—the discharge of feces. This, of course, is a natural function and refers to the removal of undigested food and other waste material from the alimentary canal through the anus; it is the end of the digestion process.

The mantra or refrain throughout this presentation has been and is to present the question: is a waste really a waste? Well, if you were to ask the majority of people if they thought what they flushed down the toilet or deposited in an outhouse or in the streets (in some locations) they would have no hesitancy in replying that absolutely human waste is waste.

But is it actually, definitively waste? Are the urine, fecal material, and other bodily discharged substances flushed down the toilet really waste?

The answer to this question and others is addressed shortly, but for now it is convenient for a clear understanding to divide human waste into excreta and urine. Wastewater treatment technology is used here to describe human excreta and urine using the industry descriptors solid- and wet-sides of human waste, respectively.

Okay, back to the present question related to whether the human waste is actually a waste or not.

DOI: 10.1201/9781003252665-12

FIGURE 12.1   A cloud of wastes by other names. Source: Adapted from K. O'Neill's *Waste* accessed @ www.politybooks.com.

To answer this question, we need to look at a bit of history and also take a look at present practice.

So let's begin by taking a look at night soil. For those not familiar with the term and its meaning, night soil is a historically used code word or euphemism for human excreta collected from gutters, outhouses, privies, cesspools, closets, middens, sinks, muckhills, heaps, pit latrines, septic tanks, and so forth. Although these depositories were and are typically described as wet and foul and most disgusting, these offensive characteristics did not and have not prevented humans from collecting, or "slopping out" (i.e., the collecting of human waste) the deposits and hauling them off, usually at night, for use in agricultural applications.

The truth be told night soil is very rich in plant nutrients. Fresh human feces contain about 1.5% nitrogen, 1.1% phosphorus, 0.5% potassium, and others. Night soil is a cheap and convenient fertilizer and is used in some locations for that purpose.

So, using human waste in the form of night soil cannot be classified as a waste. How about the wet-side of human waste? Can it be used in a useful way or manner? Answer: yes it can be reused to the benefit of us all. Consider the following section, a real-world example, of how human urine is reprocessed, made safe according to health standards and regulations, pumped into freshwater aquifers, and stored for eventual human consumption and other practical uses.

WET-SIDE OF HUMAN WASTE

MIXING NATIVE GROUNDWATER AND INJECTATE*

Ted Henifin's jaw-dropping, eyebrow-raising idea was first proposed in 2015, and last month the Hampton Roads Sanitation District [HRSD] general manager kicked off its pilot phase to stop what some scientists have called a nightmare in super slow motion.
— **Darryl Fears,** *The Washington Post***, 20 October 2016**

"To stop a nightmare in slow motion ...?" Are you kidding me? Is the real message: Shouldn't Henifin dodge the challenge or the criticism that accompanies his jaw-dropping, eyebrow-raising project? Weren't there critics who mumbled these same words and doubts when da Vinci had the audacity to be far thinking, when Newton worked his calculus, when the Wright Brothers attempted to fly like a bird, when Einstein developed his theories, when General Patton marched his army against the bad guys, and when Jonas Salk developed the vaccine for polio? Two things seem certain to me; critics abound by the mega-millions while innovators, risk takers, far thinkers, and people with grit, people with backbone and vision like Henifin are almost as rare as the Dodo Bird; moreover, it is only the innovator who thinks outside the stovepipe; the rest go up the chute, undampened.
— **Frank R. Spellman (2017)**

Bear with me as I present a very simplistic view of what this section is all about. Envision two 1-liter glass beakers. In one of these beakers, we fill it half-full of clean, safe drinking water. In the other beaker we fill it half-full of salty sea water. A normal person who wants to quench his or her thirst would obviously prefer to drink from the beaker of clean, safe water; he or she would leave the beaker of salty sea water alone. Now, if we take either the beaker of clean, safe drinking water or the beaker of salty sea water and pour one into the other we have obviously mixed the two different solutions. The question now becomes is the new 1-liter mixture of clean, safe drinking water and salty sea water something that any of us would want to drink? The truth is some would only use this mixture to gargle with. I have seen this done; have you? Anyway, the point I am making here is that by our action of mixing clean, safe drinking water with salty sea water we have changed or adulterated the contents.

HRSD does not intend to mix clean, safe drinking water with salty sea water or any other contaminant. No, the intent is not to adulterate native groundwater in any way. Instead, the intent is to inject, replenish, and recharge the Potomac Aquifer's native groundwater supply with purified, safe water from the advanced treatment of wastewater effluent. In order to do this, to ensure the injection of treated wastewater is of the same quality as the native groundwater contained in the Potomac Aquifer HRSD and its consultant (CH2M) conducted a feasibility study. This feasibility study evaluated the geochemical compatibility of recharging clean water (injectate), native groundwater, and injectate interactions with minerals in the Potomac Aquifer

---

* Information in this section is from F.R. Spellman (2021) *Sustainable Water Initiative For Tomorrow* (SWIFT).

System aquifers. Three-discrete injectate chemistries originate from advanced water treatment processes (AWTP) including reverse osmosis (RO), nanofiltration (NF), and biologically activated carbon (BAC).

The focus of the study was on a single major wastewater treatment plant where conditions (such as geography, flow, geology, injectate quality, and groundwater quality) best represent the Hampton Roads Sanitation District system. This was based on a large number of permutations involved with comparing three injectates with native groundwater chemistries from the three PAS aquifers (i.e., Upper, Middle, Lower Potomac Aquifers) beneath HRSD's seven WWTPs, and then applying the injectate chemistries to aquifer minerals in the three aquifers.

HRSD's York River Treatment Plant was selected for this evaluation and for the subsequent pilot study. In addition to displaying fairly representative conditions, the property surrounding the York River Treatment Plant is sufficiently spacious to accommodate a wastewater treatment plant (WWTP) upgraded with an Advanced Wastewater Treatment Plant (AWTP) and an injection wellfield.

HRSD'S WATER MANAGEMENT VISION

Hampton Roads region is faced with a variety of future challenges related to the management of the region's water supply and receiving water resources. These challenges involve a combination of technical, financial, and institutional complexities that invite the exploration of using non-traditional approaches that provide benefits on a larger scale beyond what the current wastewater treatment and disposal model can achieve. Accordingly, aquifer replenishment can protect and enhance the region's groundwater supplies, as well as reducing the potential damage caused by discharge (of nutrients) to the lower James River and the Chesapeake Bay and maybe slowing or arresting relative sea-level rise in the region.

PROCESSING THE WET-SIDE OF HUMAN WASTE*

After discussing the beneficial reuse potential of using human waste solids, we now turn the focus on the wet or liquid side of human waste. Simply, the question is: is human urine a beneficial reusable product? There are those few who point out that urine waste can be used as a fertilizer or in some health applications. Let's just say that in this discussion we are not heading in that direction for many reasons—primarily because usage of human urine in agriculture and medical applications is not widespread. So, instead it is the purpose of this particular discussion to direct attention toward removing the non-useful constituents (pathogens and so forth) in human urine using treatment and disinfection and then connecting the wet-side of treated human waste that has been brought up to drinking water standards to a pipe-to-pipe connection for reuse.

---

* From HRSD (2016). *Sustainable Water Initiative For Tomorrow* (SWIFT). Hampton Roads Sanitation District. Virginia Beach, VA.

Pipe-to-pipe connection?
Human urine piped to drinking water outlets?
Is this for real?
Are we supposed to go to the kitchen tap and pour a glassful of what used to
    be human urine and drink it?
Answers: Yes!

Well, the common view of such an endeavor, to use toilet flushed contents as a drinking water supply might gag the proverbial maggot, so to speak. Or such an endeavor might be referred to as foul, nasty, unpleasant, bad, horrid, dreadful, abominable, obnoxious, unsavory, repulsive, and so on and so forth. Actually, it is probably more accurate to state that the thought of drinking toilet water will and does initiate the gag factor in many people.

Okay, stepping away from the hyperbole and getting down to brass tacks, so to speak, it is probably more likely and appropriate to state that the possibility of drinking toilet water brings about or initiates the so-called yuck factor.

Yuck factor?

Yes.

But the question is why? When wastewater (toiler water) is flushed to a treatment plant and treated and then outfalled, where is it going—where is it dumped?

Answer: treated wastewater is usually outfalled (dumped) into the nearest water body—lake, lagoon, river, stream, and so forth. So, when localities withdraw water from a lake, lagoon, river, stream, and so forth they are removing treated wastewater to send to the water treatment plant to subsequently end up in your kitchen water tap. Thus, you are already drinking toilet-flushed human-deposited urine.

Let's take a closer look via Sidebar 12.1 at the so-called yuck factor involved with drinking toilet water.

SIDEBAR 12.1 WASTEWATER YUCK FACTOR OVER-STATED

That great mythical hero, Hercules, arguably the world's first environmental engineer, was ordered to perform his 5th labor by Eurystheus to clean up King Augeas' stables. Hercules, faced with a mountain of horse and cattle waste piled high in the stable area had to devise some method to dispose of the waste; he did. He diverted a couple of river streams to the inside of the stable area so that all the animal waste could simply be deposited into the river streams: Out of sight out of mind. The waste simply flowed downstream. Hercules understood the principal point in pollution control technology that is pertinent to this very day and to this discussion; that is, *dilution is the solution to pollution.*

When people say they would never drink toilet water, they have no idea what they are saying. As pointed out in my textbook, *The Science of Water*, 4th edition, the fact is we drink recycled wastewater every day. In Hampton Roads, Virginia, for example, HRSD's wastewater treatment plants outfall (discharge) treated water to the major rivers in the region. Many of the region's rivers are sources of local drinking water supplies. Even local groundwater supplies are routinely infiltrated with surface water inputs, which, again, are commonly supplied by treated wastewater (and sometimes infiltrated by raw sewage that is accidentally spilled).

My compliments to Mr. Henifin, General Manager of HRSD, who stated in a recent local newspaper article that he would be first to drink the treated wastewater effluent from the unit treatment processes at York River Treatment Plant (he did and is alive and well three years afterward). My only contention with his statement is that because of Mother Nature's Water Cycle, the one we all learned about in grade school, we have been drinking toilet water all along. I have yet to find anything yucky about it or its taste.

HRSD AND THE POTOMAC AQUIFER

A rock formation or stratum that will yield water in sufficient quantity to be of consequence as a source of supply is called an aquifer ... it is water bearing not in the sense of holding water but in the sense of carrying or conveying water.
                                        —Oscar E. Mainer (1923)

Future Generations will inherent clean waterways and be able to keep them clean.
                          —Hampton Roads Sanitation District (On-going)

When we use up one resource and then another resource and then another, there is one critical resource mankind never runs out of. What is that one item that sustains us—in a word it is innovation, innovation, innovation—innovators focus on solutions rather than problems. Because HRSD's processing of wastewater that eventually ends up in our kitchen taps and is drunk by many people in the Hampton Roads region has been introduced in this publication, it seems only correct to delve deeper into what HRSD is doing to replenish the supply of drinking water and to mitigate land subsidence in the area.

Its genesis was driven by oysters. No, not the genesis of Chesapeake Bay; its genesis was driven by a heavy, unstoppable, all knowing hand. Hampton Roads Sanitation District, arguably the premier wastewater treatment district on the globe, became a viable governor-appointed state commission monitored entity because of a significant decline in the oyster population in the Chesapeake Bay. As a case in point, consider that in the Hampton Roads region of the Chesapeake Bay in 1607 when Captain John Smith and his team settled in Jamestown, oysters up to 13 inches in size were plentiful—more than could ever be harvested and consumed by the handful of early settlers. And this population of oysters and other aquatic lifeforms remained plentiful until the population gradually increased in the Bay region.

Over-harvesting of oysters by the increased numbers of humans living in the Chesapeake Bay region was (and might still be) a major issue with the decline of the oyster population. However, the real culprit causing the decline in the oyster population is pollution. Before the Bay became polluted from sewage, sediment, and garbage disposal, oysters could handle natural pollution from stormwater runoff and other sources. Ninety years ago, when there was a much larger oyster population than today, it is estimated that the large oyster population could filter pollutants from the Bay and clean it in as little as four days. By the 1930s, however, the declining oyster population was overwhelmed by the increasing pollution levels.

For years, and in many written accounts, the author has stated that pollution is a judgment call. That is, pollution, as viewed by one person, may not be pollution observed by another. You might shake your head and ask a couple of questions, "Pollution is a judgment call? Why is pollution a judgment call?" A judgment is based on an opinion; it is an opinion because people differ in what they consider to be a pollutant based on their assessment of the accompanying benefits and/or risks to their health and economic well-being posed by the pollutant. For example, visible and invisible chemicals spewed into the air or water by an industrial facility might be harmful to people and other forms of life living nearby. However, if the facility is required to install expensive pollution controls, forcing the industrial facility to shut down or move away, workers who would lose their jobs and merchants who would lose their livelihoods might feel that the risks from polluted air and water are minor weighed against the benefits of profitable employment and business opportunity. The same level of pollution can also affect two people quite differently. Some forms of air pollution, for example, might cause a slight irritation to a healthy person but cause life-threatening problems to someone with chronic obstructive pulmonary disease (COPD) like emphysema. Differing priorities lead to differing perceptions of pollution (concern at the level of pesticides in foodstuffs generating the need for wholesale banning of insecticides is unlikely to help the starving). No one wants to hear that cleaning up the environment is going to have a negative impact on them. The fact is public perception lags behind reality because the reality is sometimes unbearable.

Well, it can be said that every problem has a solution. And that has proven to be true in most cases, with many more worldwide problems that need solutions to be found, hopefully. With regard to the problems with the Chesapeake Bay, land subsidence, diminishing groundwater, and relative sea-level rise in the Hampton Roads region, HRSD has developed the innovative Sustainable Water Initiative For Tomorrow (SWIFT)* program (a work in progress; a decadal project). Do not confuse the acronym swift with the adjectives for fast, speedy, rapid, hurried, immediate, or quick. SWIFT is a long-term project that is being developed on a timeline that is set for the installation of the technical equipment and operational procedures with a completion date of 2030.

What is HRSD's SWIFT? SWIFT is a program to inject treated wastewater into the subsurface; specifically, it is designed to inject treated wastewater to drinking water quality into the Potomac Aquifer. Injection of water into the subsurface is expected to raise groundwater pressures, thereby potentially expanding the aquifer system, raising the land surface, and counteracting land subsidence occurring in the

---

* Much of the information in this section is based on HRSD's *Proposal to establish and extensometer station at the Nansemond wastewater treatment plant in Suffolk, VA* (2016). Proposed by USGS.

Virginia Coastal Plain. In 2016, a pilot project site is under construction at the HRSD Nansemond Wastewater Treatment Plant in Suffolk, Virginia, to test injection into the aquifer system. HRSD has asked the United States Geological Survey (USGS) to prepare a proposal for the installation of an extensometer monitoring station at the test site to monitor groundwater levels and aquifer compaction and expansion.

PROBLEM

SWIFT is designed to counter land subsidence at various locations in the Hampton Roads area of southern Chesapeake Bay where land subsidence rates of 1.1 to 4.8 millimeters per year have been observed (Eggleston and pope, 2013; Holdahl and Morrison, 1974).

Injection of treated wastewater (treated to drinking water quality) is expected to counteract land subsidence or raise land surface elevations in the region. Careful monitoring of aquifer-system compaction and groundwater levels can be used to optimize the injection process and to improve fundamental understanding of the relation between groundwater pressures and aquifer-system compaction and expansion.

There is more to HRSD's treated wastewater injection project, SWIFT, than just arresting or mitigating land subsidence and relative sea-level rise in the Hampton Roads region. One of the additional goals of the project is to stop the discharge of treated wastewater from seven of its plants. That would mean 18 million pounds a year less of nitrogen, phosphorus, and sediment out-falling into the bay. Assuming SWIFT works as designed, this is a huge benefit to the Chesapeake Bay in that it may help to prevent or reduce the formation of algal bloom dead zones. Not only would success as a result of treated wastewater injection benefit the bay, but it would also be a huge benefit for the ratepayers at HRSD. To meet regulatory guidelines to remove nutrients from the discharged treated wastewater would cost hundreds and millions of dollars and almost non-stop retrofitting at the treatment plants to keep up with advances in treatment technology and regulatory requirements. Another goal of HRSD's SWIFT project is to restore or restock potable groundwater supplies in the local aquifers. The drawdown of water from the groundwater supply has not only contributed to land subsidence but to a reduction of water available for potable use.

HRSD's planned restocking Hampton Roads groundwater supply with injected wastewater treated to potable water quality is not without its critics. The critics state that HRSD's wastewater injection project would contaminate potable water aquifers. For the critics and others this is where the so-called "yuck factor" comes into play. The yuck factor, in this particular instance, has to do with the thought that groundwater for consumptive use will be contaminated basically with toilet water. This is the common view of many of the critics who feel HRSD's SWIFT project is nothing more than direct reuse of wastewater; that is, a pipe-to-pipe connection of toilet water to their home water taps.

What the critics and others do not realize is that we are already using and drinking treated and recycled toilet water. As far as HRSD's SWIFT project contaminating existing aquifers with toilet water, it is important to point out that this water is to be treated (and already is at Nansemond Treatment Plant in Suffolk, Virginia) to drinking water quality—to drinking water quality is the key phase here. This sophis-ticated and extensive train of unit drinking water quality treatment processes—treated wastewater that the HRSD General Manager and several others drank right out of the process recently, and, by the way, they are doing just fine today, thank you very much—is discussed in detail later in the text. The bottom line: Statements about the yuck factor involved in drinking treated toilet water are grossly over-stated, as pointed out in Sidebar 12.1.

HRSD's SWIFT project proposes to add an advanced treatment process to several of its facilities to produce water that exceeds drinking water standards, and to pump this clean water into the ground and the Potomac Aquifer. This will ensure a sus-tainable source of water to meet current and future groundwater needs throughout eastern Virginia while improving water quality in local rivers and the Chesapeake Bay. SWIFT project benefits include (HRSD, 2016):

- Eliminates HRSD discharge to the James, York, and Elizabeth rivers except during significant storms.
- Restores the rapidly dwindling groundwater supplies in eastern Virginia upon which hundreds of thousands of Virginia residents and businesses depend.
- Creates huge reductions in the discharge of nutrients, suspended solids, and other pollutants to the Chesapeake Bay.
- Makes available significant allocations of nitrogen and phosphorous to sup-port regional needs.
- Protects groundwater from saltwater contamination/intrusion.
- Reduces the rate of land subsidence, effectively slowing the rate of sea-level rise by up to 25%.
- Extends the life of protective wetlands and valuable developed low-lying lands.

POTOMAC AQUIFER

Before HRSD commenced pumping up to 130 million gallons per day (MGD) into the Potomac Aquifer System or as it will do more in the future into any other aquifer—along with the expert assistance of CH2M, its primary consultant in this matter—HRSD first had to determine the feasibility of aquifer replenishment by recharging clean water, purified from the advanced treatment of wastewater treat-ment plant (WWTP) effluent. In this section, a description of the essential elements of recharging clean water into the Potomac Aquifer System (PAS) at the following seven Hampton Roads Sanitation District WWTPs: Army Base, Boat Harbor, James River, Nansemond, Virginia Initiative Plant, Williamsburg, and York River. Also, a determination of the capacity of individual injection wells at the seven WWTPs is made or is being studied at the present time; moreover, a projection of the injection capacity within the existing site area of the seven WWTPs has or is being deter-mined; and a characterization of the regional beneficial hydraulic response of the PAS to clean water injection has and is being determined.

The material presented in this section is based on data available in city/country, state, and federal databases, reports, scientific papers, interviews, operator findings, and other literature to characterize the injection capacity of individual wells at each of the WWTPs. Injection well capacities and analytical mathematical modeling were used to estimate the injection capacity of each WWTP based on the plant's flow rate, property size, and the transmissivity of the underlying PAS aquifer.

The Potomac Formation*

Given the elevated volume requiring disposal, and the importance of minimizing the number of injection wells, the most suitable aquifer units are those that exhibit the highest production capacity. Furthermore, a thick, confining bed composed of impermeable materials like silt or clay should overlie the aquifer to prevent vertical migration of the injection fluid (injectate) into the surrounding aquifer units. Beneath the HRSD service area, the Cretaceous age, Potomac Formation meets these criteria. The Potomac Formation contains thick sand deposits, forming three-discrete aquifer units. Although the modern convention developed by the USGS and the Virginia Department of Environmental Quality (VDEQ) is to group the three aquifers as one, named the Potomac Aquifer (McFarland and Bruce, 2006), locally and in this book, because they behave hydraulically as three distinct units, they must be and are examined separately. Accordingly, in this presentation these units are treated as three distinct units referred to as Upper Potomac Aquifer zone (UPA), Middle Potomac Aquifer zone (MPA), and Lower Potomac Aquifer zone (LPA) (Laconia and Meng, 1988; Hamilton and Larson, 1988). Each discrete aquifer in the HRSD service area is separated from adjacent aquifers by clay confining beds of measurable thickness, while a cumulative thickness of silt and clay units totaling several hundred feet overlies the top (UPA) aquifer unit of the PAS. Production wells screened in the PAS exhibit significantly greater pumping capacities than wells screened in other aquifers of the Virginia Coastal Plain (Smith, 1999). Some wells can pump at rates approaching 3,000 gallons per minute (gpm) (4.3 MGD). In addition to confinement and production capacity, aquifers in the PAS exhibit deep static water levels, ranging from 80 to 180 feet below grade (fbg), providing available head for injection.

Recent USGS sedimentological studies suggest the HRSD service area, spanning the York–James Peninsula and Southeastern Virginia is well situated regarding the quality of aquifers in the PAS (McFarland, 2013). PAS aquifer sands display greater thickness, coarser grain size, and better sorting in the HRSD service area than units in the northern Virginia Coastal Plain, or to the south in the northern North Carolina. As a result, aquifers exhibit excellent hydrologic coefficients (hydraulic conductivity, transmissivity), and thus, more productive well capacity (HRSD, 2016).

The Potomac Aquifer System outcrops at the ground surface west of King William WWTP in King William County Virginia. PAS is recharged by infiltrating precipitation. In the recharge area, the aquifers range in thickness from 70 (UPA) to

---

* Much of the information in this section is based on information from HRSD (2016) and compiled by CH2M.

400 feet (LPA). Further downdip, recharge enters the PAS by leakage through overlying confining beds (Meng and Harsh, 1988).

Individual aquifers thicken and dip to the southeast toward the Atlantic coastline reaching thickness ranging from 170 feet (UPA) to approaching 1,000 feet (LPA) at the coast (Treifke, 1973). These thicknesses comprise all sediments contained in the vertical section including discrete sand beds representing aquifer materials, and interleaving silt and clay lenses from intra-aquifer contained beds. Individual aquifers exhibit a strongly interbedded morphology consisting of thin to thick beds of sands, silts, and clays. Beneath Newport News, the MPA consists of six discrete sand intervals. Obtaining maximum production (or injection) capacities from wells installed in layered aquifers such as the PAS requires extending the well screen assembly across the maximum thickness of aquifer sand. Screen assemblies can consist of multiple screens and blank sections (HRSD, 2016).

Injection Wells

Typically, when we think about a well, we have the image of a hole in the ground and some type of device to pump water to the surface for use or storage. A well pumping water from the subsurface to the surface functions to provide whatever use it might be intended for. However, wells used for pumping are not what we are concerned with in this book. We are concerned with just the opposite; that is, wells that inject treated injectate and that do not pump fluids to the surface.

Injection wells are known as Class V wells. There are 22 types of Class V injection wells. The types of wells are shown in Table 12.1.

Subsidence Control Wells

It is the last type of Class V wells listed in Table 12.1 (subsidence control wells) that is the focus of this presentation. Subsidence control wells are injection wells whose primary objective is to reduce or eliminate the loss of land surface elevation due to removal of groundwater providing subsurface support. Subsidence control wells are important to HRSD's SWIFT project. The goal is to inject treated wastewater to drinking water quality into the underground Potomac Aquifer System to maintain fluid pressure and avoid compaction and to ensure that there is no cross-contamination between the infected water and underground sources of drinking water. Thus, the injectate must be of the same quality, or superior in quality as the existing groundwater supply.

Injection Well Hydraulics

With regard to aquifer injection hydraulics, a well's injection capacity depends on its specific capacity and the pressure (head) available for injection, a function of the static head (water level) of the aquifer in which the well is screened. Specific capacity describes a well's yield per unit of head decrease (drawdown) in a pumping well, or head increase (draw-up) in an injection well. Specific capacity is expressed in units of feet of drawdown/draw-up per unit of pumping/injection rate in gpm per foot (gpm/ft), respectively. When injection begins, the water level in the well rises as a function of the transmitting properties of the receiving aquifer and the well's

TABLE 12.1

Class V Underground Injection Wells

Type of Injection Well	Purpose
Agricultural Drainage Wells	Receive agricultural runoff
Stormwater Drainage Wells	Dispose of rainwater and melted snow
Carwashes Without Undercarriage Washing or Engine Cleaning	Dispose of wash water from car exteriors
Large-Capacity Septic Systems	Dispose of sanitary waste through a septic system
Food Processing Disposal Wells	Dispose of food preparation wastewater
Sewage Treatment Effluent Wells	Used to inject treated or untreated wastewater
Laundromats Without Dry Cleaning Facilities	Dispose of fluid from laundromats
Spent Brine Return Flow Wells	Dispose of spent brine for mineral extraction
Mine Backfill Wells	Dispose of mining byproducts
Aquaculture Wells	Dispose of water used for aquatic sea life cultivation
Solution Mining Wells	Dispose of leaching solutions (lixiviants)
In-Situ Fossil Fuel Recovery Wells	Inject water, air, oxygen solvents, combustibles, or explosives into underground or oil shale beds to free fossil fuels
Special Drainage Wells	Potable water overflow wells and swimming pool drainage
Experimental Wells	Used to test new technologies
Aquifer Remediation Wells	Use to clean up, treat, or prevent contamination of underground sources of drinking water
Geothermal Electrical Power Wells	Dispose of geothermal fluids
Geothermal Direct Heat Return Flow Wells	Dispose of spent geothermal fluids
Heat Pump/Air Conditioning Return Flow Wells	Re-inject groundwater that has passed through a heat exchanger to heat or cool buildings
Saline Intrusion Barrier Wells	Injected fluids to prevent the intrusion of salt water
Aquifer Recharge/Recovery Wells	Used to recharge an aquifer
Noncontact Cooling Water Wells	Used to inject noncontact cooling water
Subsidence Control Wells	Used to control land subsidence caused by groundwater withdrawal or over pumping of oil and gas

efficiency (Warner and Lehr, 1981). While the transmitting character of the aquifer should remain stable over the service life of an injection well, the available head for injection will decline as injection recharges the aquifer causing the static water level to rise toward the ground surface (HRSD, 2016).

In this discussion of the planning and study phase (and now the processing phase at Nansemond Treatment Plant, NATP) of the SWIFT project, it is important to differentiate between production and injection by referring to injection-specific capacity as injectivity. Moreover, for our purposes here, the evaluation of the specific capacity of local production wells and its conversion to injectivity forms an important variable in determining the capacity of individual injection wells and ultimately the total injection capacity across the affected area, and the number of wells required at each WWTP.

DID YOU KNOW?

Specific capacity is one of the most important concepts in well operation and testing. The calculation should be made frequently in the monitoring of well operation. A sudden drop in specific capacity indicates problems such as pump malfunction, screen plugging, or other problems that can be serious. Such problems should be identified and corrected as soon as possible. *Specific capacity* is the pumping rate per foot of Drawdown (gpm/ft), or

$$\text{Specific capacity} = \text{Well yield} \div \text{drawdown}$$

Problem: If the well yield is 300 gpm and the drawdown is measured to be 20 ft, what is the specific capacity?

Solution:

$$\text{Specific Capacity} = 300 \div 20$$

$$\text{Specific Capacity} = 15 \text{ gpm per ft of drawdown}$$

In determining the injectivity for wells screened in specific aquifer units, specific capacity values were evaluated in the three Potomac Aquifer units with respect to their proximity to HRSD's wastewater treatment plants (WWTPs). The VDEQ supplied a database of production wells that contained a total of 98 wells screened in either the UPA, MPA, or LPA, with some wells screened across sands in two of the three aquifers. In these wells, a technique was employed to determine the specific capacity contributed by each aquifer to the well, based on the length of the screen penetrating the aquifer, divided by the total length of the well screen. It proved necessary to separate the aquifers across individual wells for the less represented LPA. Because of its greater depth and water quality, production wells rarely penetrated all the sand units in the LPA, often screening sand intervals in both the MPA and UPA.

Across the study area, the specific capacity of production wells screening the UPA and MPA averaged 35.5 and 32.4 gpm/ft respectively, essentially equaling each other, while wells screened in the LPA exhibited a 40% lower value of 21 gpm/ ft. Wells screening the LPA, as identified by VDEQ, were only located around the Franklin Paper Mill and Franklin City area. Wells screened in the UPA and MPA were better represented across the study area.

As stated previously, the available head for injection represents an important factor in determining injection capacities. Multiplying the well's injectivity by the available head for injection provides the injection capacity of the well. As a driver for this study, local industrial and residential development has resulted in elevated pumpage, drawing water levels in aquifers of the Potomac Aquifer System downward at rates averaging 1 to 2 feet per year (McFarland et al., 2006). The declining water levels represent a greater available head for injection. However, once injection commences, water levels should rebound with a corresponding loss in available head (HRSD, 2016).

Injection Operations

Because of the deep static water levels at most HRSD wastewater treatment plants and available head for injection in feet at each site for each level of the Potomac Aquifer (see Table 12.2), HRSD could inject water under gravity conditions or

TABLE 12.2

Summary of Available Head for Injection (ft) in Potomac Aquifer System at HRSD's Treatment Plants

Wastewater Treatment Plant	Ground Elevation	PAS Aquifer	Available Head for Injection (Ft)
Army Base	11	UPA	107.28
		MPA	101.28
		LPA	101.28
Virginia Initiative Plant	8.5	UPA	109.78
		MPA	106.28
		LPA	101.28
Nansemond	22.5	UPA	123.78
		MPA	106.28
		LPA	106.28
Boat Harbor	4	UPA	100.28
		MPA	91.28
		LPA	96.28
James River	18	UPA	107.28
		MPA	81.28
		LPA	86.28
York River	5.5	UPA	86.78
		MPA	76.28
		LPA	76.28
Williamsburg	58.5	UPA	125.78
		MPA	67.28
		LPA	76.28
Rate of water-level decline	0.92ft/yr		

Note:

ft = feet

Source: Adapted from CHM2, *Sustainable Water Recycling Initiative: Groundwater Injection Hydraulic Feasibility Evaluation*, Report No. 1, CH2M, Newport News, VA, 2016.

pressurized conditions. Under gravity conditions, HRSD would fit a foot valve to the base of the injection piping to reduce head and maintain a positive pressure in the injection and well-header piping. This design facilitates better control of injection rates compared with cascading water down the injection piping, inducing a vacuum through the system. If HRSD decides to inject under gravity conditions, it would not be necessary to seal the injection head, well-header piping, and associated fittings. Instead, the design requires monitoring injection levels rising in the well's annular space to prevent it from topping the ground surface.

DID YOU KNOW?

"No owner or operator shall construct, operate, maintain, convert, plug, abandon, or conduct any other injection activity in a manner that allows the movement of fluid containing any contaminant into underground sources of drinking water, if the presence of that contamination may cause a violation of any primary drinking water regulation under 40 CFR part 142 or may otherwise adversely affect the health of persons." (40 CFR 144.121)

An option to injecting under gravity conditions could entail sealing the injection head, well-header piping, and associated fittings, while allowing the injection head to rise above the elevation of the ground surface by maintaining a positive pressure in the annular space of the well. With a greater available head for injection, HRSD could achieve higher injection rates, while anticipating risking regional water levels inherent with injecting large volumes of water.

The operator is responsible to ensure annular pressures stay below an established threshold. Elevated annular pressure can stress the sand filter pack surrounding the well screen by lifting it and initiating the formation of channels that connect the well screen to the surrounding formation materials. Limiting the injection pressure to 10 pounds per square inch (psi) in the well's annular space precludes damage to the filter pack while making available another 23 feet of head for injection.

Regular maintenance is required to operate injection wells to their maximum capacity whenever they are screening fine sandy materials like those found in the Potomac Aquifer System. In the Aquifer Storage and Recovery (ASR) wells, screening the Potomac Aquifer System, of similar aquifers at Chesapeake, Virginia, and in North Carolina, Delaware, and New Jersey, fine sandy aquifers have proven susceptible to clogging from total suspended solids (TSS) entrained in the recharge water in the recharge water (McGill and Lucas, 2009). Even high-quality treated water can contain some amount of TSS that accumulates (clog) in the screen and pore spaces of sand filter pack and formation. Clogging reduces the permeability around the screen, filter pack, and aquifer proximal to the well (wellbore environment), resulting in higher injection levels while reducing injection capacity.

Aquifer storage and recovery wells operating in these states are equipped with conventional well pumps, allowing periodic well back-flushing during recharge. Back-flushing entails temporarily shutting down recharge and turning on the well pump for a sufficient time to remove fine-grained materials from the wellbore environment to the ground surface.

By adopting the approach in maintaining ASR wells by installing a pump in each of HRSD's injection wells for back-flushing, it will slow the accumulation of fine-grained materials in the wellbore environment. To generate sufficient energy required to effectively remove solids from the wellbore, the capacities of back-flushing pumps should equal or exceed the injection rates planned for the well.

DID YOU KNOW?

The type and quality of injectate and the geology affect the potential for endangering an underground supply of drinking water. The following examples illustrate potential concerns.

If injectate is not disinfected, pathogens may enter an aquifer. Some states allow injection of raw water and treated effluent. In these states, the fate of microbes and viruses in an aquifer is relevant.

When water is disinfected prior to injection, disinfection byproducts can form in situ. Soluble organic carbon should be removed from the injectate before disinfection. If not, a chlorinated disinfectant may react with the carbon to form contaminating compounds. These contaminants include trihalomethanes and haloacetic acids.

Chemical differences between the injectate and the receiving aquifer may create increased health risks when arsenic and radionuclides in the geologic matrix interact with injectate having a high reduction-oxidation potential.

Carbonate precipitation in carbonate aquifers can clog wells when the injectate is not sufficiently acidic (USEPA, 2016).

Even with back-flushing, progressive clogging can occur and exceed the ability of back-flushing to maintain injection wells near their maximum capacity. Each WWTP should contain a sufficient number of wells to compensate for removing wells from service for rehabilitation, without compromising the facility's injection capacity.

INJECTION WELL CAPACITY ESTIMATION

In the planning and study stage, the following steps were employed in estimating the capacities of injection well in the Potomac Aquifer System at each HRSD wastewater treatment plant (HRSD, 2016):

- Estimate specific capacity at individual wells (Specific capacity = well yield ÷ drawdown).
- Organize specific capacities by aquifer.
- Separate well screens spanning two aquifers and calculate specific capacity for both.

- In short screen assemblies, normalize screen length to 100 feet.
- Convert specific capacity to injectivity.
- Calculate available head for injection from USGS 2005 synoptic study.
- Combine injectives of UPA and MPA, or MPA and LPA in a single injection well.
- Average available head for injection across UPA and MPA, or MPA and LPA in a single injection well.
- Add 23 feet of head to available injection head to account for maintaining a pressure of 10 psi in annular space of well.
- Multiply injectivity by available head for pumping to obtain injection well capacity.
- Practically limit injection well capacity to 3 MGD (2,100 gpm).
- To estimate the number of injection wells per WWTP, divide the plant's effluent rate by injection well capacity.
- To facilitate periodic maintenance, add one injection well for every five, per WWTP.

ESTIMATING SPECIFIC CAPACITY AND INJECTIVITY

The VDEQ database of wells and their locations were obtained and evaluated according to their location relative to HRSD's WWTPs at Army Base, Boat Harbor, James River, Nansemond, Virginia Initiative Plant, Williamsburg, and York River. The database contained static water levels, pumping levels, and stable production rates at most wells, enabling the calculation of specific capacity. Moreover, the VDEQ database identified the Potomac Aquifer System unit spanned by each screen interval in each well. Several wells featured over five well screen intervals. As previously described, maximizing well screen length increases a well's production capacity. Accordingly, many larger-capacity productized wells were equipped with multiple screen intervals, and in many cases, screening more than one Potomac Aquifer System unit (HRSD, 2016).

Specific capacities were calculated for each well and then grouped according to the aquifer unit(s) in which the well was screened. Specific capacities for wells with screens spanning two aquifers resulted in developing two specific capacities for a single well. Screen assemblies in these wells usually spanned the UPA and MPA, or MPA and LPA. Individual intervals were grouped according to the aquifer which they screened. In wells spanning several aquifers, the specific capacity for a discrete aquifer was then estimated by taking the length of screen spanning the aquifer and dividing it by the total screen length as follows:

$$\text{Total SC of well} \times \left(\text{SL Aquifer 1/TSL of well}\right) = \text{SC Aquifer 1}$$

where
SC = specific capacity
SL = screen length
TSL = total screen length

HRSD should design their injection wells with the screens penetrating an entire aquifer's sand thickness to maximize injection capacity. Note, however, that many of the production wells studied in this project used shorter screen assemblies that only partially penetrated the aquifers in which they were installed. Accordingly, to make the estimates compatible with fully penetrating injection wells, the specific capacity for a shortened screen assembly was normalized to a well with 100 feet of screen by applying the following equation.

SC Aquifer 1 = (100 feet of screen/SL) = SC of aquifer normalized to 100 feet of screen.

One hundred feet of screen was recognized as an average total for assemblies fully penetrating the UPA, MPA, and LPA. Specific capacities for wells exhibiting screens extending over 100 feet across a specific aquifer unit of the PAS were accepted as representative and applied without modification (HRSD, 2016).

Based on the observations from aquifer storage and recovery wells, which function in both injection and pumping modes of operation, the specific capacity and the injectivity of wells installed in the same aquifer usually vary slightly (Pyne, 2005). In a production well, water migrates toward a potentiometric head lowered by pumping in the wellbore. Water moves through an environment constrained by the size and heterogeneity of the porous media into the well. Upon entering the pumping well, the water is no longer impeded by porous media.

By comparison, injected water is driven down the wellbore by an elevated head against the resistance of the wellbore environment. As a result of the greater resistance to flow, the injectivity of an ASR well often falls 10–50% less than its specific capacity (Pyne, 2005). To re-create this important relationship for the study, converting specific capacity to injectivity involved multiplying specific capacity by a factor of 0.5, an average of the ratio of injectivity to specific capacity in Atlantic Coastal Plain injection type wells (Pyne, 1995).

AVAILABLE HEAD FOR INJECTION

The available head for injection also played an important role in determining the capacity of an injection well. The available head in each aquifer at individual WWTPs was determined by obtaining the most recent general view of the whole (synoptic) water level information. The US Geodetic Survey last measured synoptic water levels separated by aquifer unit of the Potomac Aquifer System, in 2005 (VADEQ, 2006b). More recent work considers all aquifers of the Potomac Aquifer System together.

Note that with progressively declining water levels attributed to over pumping, potentiometric heads measured in 2005 cannot accurately calculate septic conditions in 2014. To adjust potentiometric levels to 2015 conditions, hydrographs were examined to quantify the annual conditions in water levels (VADEQ, 2006a). The evaluation revealed that potentiometric levels have declined an average of 0.9 feet per year since 2005 in the Upper Potomac Aquifer, Middle Potomac Aquifer, and Lower Potomac Aquifer. The annual rate of decline in potentiometric levels was multiplied by 10 years and then applied to the potentiometric head from 2005, projecting elevations to 2014.

In estimating the depth of water, the projected potentiometric level for 2014 in each aquifer unit was subtracted from the elevation of the land surface at each WWTP. A constant head of 23 ft, equal to injecting under a pressure of 10 psi in the well's annual space, was added to the depth of the water level to obtain the available head for injection (HRSD, 2016).

Flexibility for Adjusting Injection Well Capacities

Maximizing the capacity of individual injection wells required screening two aquifer units, either the Upper Potomac Aquifer and Middle Potomac Aquifer, or the Middle Potomac Aquifer and Lower Potomac Aquifer in each well. Because of potential hydraulic inefficiencies inherent with the difference in screen elevations, wells screening the Upper Potomac Aquifer and Lower Potomac Aquifer were not considered for this evaluation. This is important because it provides HRSD with some flexibility for adjusting injection well capacities while installing the injection well-fields. Upon installing the initial injection wells at a WWTP, HRSD could elect to screen all three Potomac Aquifer System units in a single well or revert to screening one aquifer if capacities fail to meet, or significantly exceed expectations, respectively.

To obtain the injection well capacity, the final steps in estimating injection well capacities were determined by adding the injectivities of two aquifer units in a well and averaging the available head for injection between the two aquifers. Then, the resulting injectivity was multiplied by the averaged available head for injection.

NUMBER OF INJECTION WELLS REQUIRED AT EACH WASTEWATER TREATMENT PLANT

Elevated well capacities that exceeded reasonable constructability practices resulted from the contamination of high injectivities from merging screens and large available heads for injection. Constructing the relatively deep wells of sufficient diameter to inject at rates approaching 8 MGD (5,600 gpm) or equipping the well with a back-flush pump capable of the same rate are likely impractical. To reduce well casing, screen, and pumps to dimensions consistent with local well drilling capabilities and operation (such as equipment available and electrical service), a capacity threshold of 3 MGD was set for the injection's wells.

The number of injection wells at each WWTP was determined by dividing the plant's effluent rate by the injection well capacities. At large plants, injection wells were split evenly between UPA and MPA, and MPA and LPA combinations. These well totals and aquifer combinations at each WWTP formed the basis for the initial mathematical modeling runs discussed in the modeling section of this work. To accommodate removing wells from service for rehabilitation, one well was added for each five. At plants with smaller effluent flows, one well was added to any total less than five.

HRSD's Virginia Initiative Plant (VIP) and Nansemond Treatment Plant required the largest number of injection wells to replenish the aquifer because of their higher effluent rates and relatively low injectivity in the Middle Potomac Aquifer. Lower injectives also appear in the Upper Potomac Aquifer adjacent to the Virginia Initiative Plant.

Production wells supporting the mapping of specific capacity in the Middle Potomac Aquifer for this project were not present within a 5-mile radius of the Nansemond and Boat Harbor Plants. Regionally, specific capacity values in the Middle Potomac Aquifer appear to increase to the southeast with the exception of two production wells located west of Nansemond and Boar Harbor. To maintain a conservative approach to the project, low-specific capacity values imparted by these wells were maintained in estimating injection well capacity at Nansemond and Board Harbor. By comparison, large-specific capacity values in the Upper Potomac Aquifer and Middle Potomac Aquifer around the James River, York River, and Williamsburg Plants resulted in elevated injectivities, which yielded elevated hypothetical injection capacities despite the relatively shallow available head for injection (HRSD, 2016).

AQUIFER INJECTION MODELING

Hydrologists, hydrogeologists, and groundwater experts in other professional fields soon learn that groundwater flow models are simplified representations of often highly complex hydrogeological flow systems. Modeling tools are well suited for analyzing aquifer injection experiments. Generally, incorporating as much available hydrogeologic information as possible into the formulation of the conceptual and numerical models of the flow system is advantageous. This is the approach used by HRSD and its consultant in modeling for HRSD's SWIFT project. Hydrogeologic information takes many forms, including maps that show outcropping surfaces of geologic units and faults, cross sections derived from geophysical surveys and wellbore information that show the likely subsurface location of geologic units and faults, maps of water-table levels, independent point well data, and maps showing the hydraulic properties of the subsurface materials. HRSD and its consultant used this information to classify the geologic units into hydrogeologic units, which are convenient units with which to define hydrologic properties (Anderman and Hill, 2000). The modeling employed in the SWIFT project helps to project or estimate the capacity of injections of injectate and other important parameters used in this project.

Estimating the capacity of individual injections at each WWTP along with a preliminary determination of the number of wells at each facility comprises one of several key elements of HRSD's SWIFT; the significance of these determinations can be seen when the goal is to dispose of nearly 130 MGD into the Potomac Aquifer System. The modeling executed in this section tests the hydraulic interference between injection wells located within the boundaries of each WWTP property. This evaluation will identify whether the individual WWTP properties are sufficiently large to contain the projected number of wells required to dispose of the projected effluent volumes.

MATHEMATICAL MODELING

Estimates of injection well capacity and the appropriate number of wells assigned at each WWTP described earlier did not account for hydraulic interference between wells in the same aquifer. With well screens combining the UPA and

MPA or MPA and LPA, hydraulic interference in the MPA will exert the greatest influence on local injection levels. Wastewater treatment plants that are situated on smaller properties will cause an increase in hydraulic interference; they require smaller inter-well spacing for fitting the number of injection wells required to dispose of effluent.

Mathematical modeling techniques were used in quantifying the interference between injection wells located at the WWTPs and rebounding water levels in the aquifers receiving effluent. In this section, analytical groundwater flow modeling is used to evaluate local groundwater mounding at individual WWTPs while injecting effluent (injectate) over 50 years.

DID YOU KNOW?

Hydraulic engineers and others are quite familiar with mathematical modeling. With the continuing advancements in computer technology and development of advanced computer engineering programs, engineers rely more and more on mathematical modeling. A mathematical model is an abstract model that uses mathematical language to describe the behavior of a system. Eykhoff (1974) defined a mathematical model as a representation of the essential aspects of an existing system (or a system to be constructed) which presents knowledge of that system in usable form.

GROUNDWATER FLOW MODELING

In evaluating potentiometric levels in the Upper Potomac Aquifer, Middle Potomac Aquifer, and Lower Potomac Aquifer, analytical groundwater flow modeling was applied at each injection well at the seven WWTPs. The modeling study extends the determination of individual injection well capacities by testing the injection rates under the spatial conditions unique to each WWTP property. Although the injection capacities of individual wells may appear feasible at a WWTP based on the head available for injection and the aquifer transmissivities, hydraulic interference between multiple wells can drive injection levels higher in inverse proportion to the available spacing between wells.

At smaller WWTP sites, interfering wells could cause injection levels to exceed 23 feet above the ground surface, the maximum threshold, established for injection head at individual wells. Mitigating elevated injection levels can entail screening all three PAS in a single well and/or reducing injection rates sufficiently to lower levels below the site injection elevation threshold. Lowering injection rates so that heads fall below site thresholds effectively limits flow injection rates lower than the projected 2040 target flows.

Accordingly, groundwater flow models were customized according to property size and the projected number of wells required at each WWTP. The computer program CAPZONE (Bair, Springer, and Roadcap, 1992) was applied to conduct the analytical groundwater flow modeling at each WWTP.

CAPZONE is an analytical flow model that can be used to construct groundwater flow models of two-dimensional flow systems characterized by isotropic and homogeneous confined, leaky-confined, or unconfined flow conditions. CAPZONE computes drawdowns at the intersections of a regularly spaced rectangular grid produced by up to 100 wells using either Equation 12.1 for a confined aquifer developed in 1935 (while working for USGS) or the Hantush–Jacob equation for a leaky-confined aquifer (see Equation 12.2) (Bair, Springer, and Roadcap, 1992). Unlike the numerical mathematical techniques employed by models like MODFLOW (which comprises simple algebraic equations that a computer cycles through multiple iterations to solve the flow equation), CAPZONE directly solves the differential flow equation. Subsequently, CAPZONE provides a more exact and conservative solution that models relying on numerical methods. However, analytical groundwater flow models offer less flexibility in simulating the heterogeneous conditions exhibited in natural systems, including multiple layers, variable grid spacing, spatially varying transmissivity, and boundary conditions. At the scale of a single WWTP, where neither the USGS nor Department of Environmental Quality (DEQ) have characterized heterogeneity beyond single wells, CAPZONE offers a reasonable method for simulating the hydraulic response to injection (or pumping) in the PAS.

The Theis Equation is simply

$$s = \frac{Q}{4\pi T} W(u) \tag{12.1}$$

$$u = \frac{r^2 S}{4 T_t}$$

where
s = drawdown (change in hydraulic head at a point since the beginning of the test)
u = a dimensionless time parameter
Q = the discharge (pumping) rate of the well
T and S = are the transmissivity and storativity of the aquifer around the well
r = distance from the pumping well to the point where the drawdown was observed
t = the time since pumping began (seconds)
W(u) = well function

The Hantush–Jacob well function for leaky-confined aquifers is abbreviated $w(u, r/B)$. The Hantush–Jacob equation can be written in compact notation as follows:

$$s = \frac{Q}{4\pi T} W(u, r/B) \tag{12.2}$$

where
s = drawdown
Q = pumping rate
T = transmissivity

In practice, CAPZONE produces drawdowns/draw ups that are then subtracted from water levels that form either a uniform or non-uniform hydraulic gradient. Thus, the analyst can designate a hypothetical gradient of one based on an observed water-level distribution (non-uniform). The non-uniform option has proven particularly useful for injection wellfield analyses at sites where a potentiometric surface exhibits irregularities or deflections that could potentially alter the potentiometric surface geometry.

At HRSD's individual WWTPs, a uniformed hydraulic gradient representing the site's position in the regional potentiometric surface for the PAS developed by USGS was input to CAPZONE. In this approach, the regional gradient was considered locally at each WWTP in estimating ambient groundwater flow direction and hydraulic gradient.

In applying CAPZONE, the boundaries of the simulated area were defined, and then the area was divided into a grid. The grid and cell dimensions for CAPZONE were unique to each WWTP, depending on the size of the site. The grid contained up to 75 columns and 75 rows, with a grid node spacing range from 11 to 200 ft at the Boat Harbor and York River WWTPs, respectively.

Other inputs included coefficients of transmissivity from the results of aquifer testing conducted at production wells in the vicinity of each WWTP and a storage coefficient of 0.0001, typical for a confined aquifer. Hydraulic gradients averaged around 0.0001 feet per year foot (ft/ft), with varying directions of groundwater flow based on the site's position within the USGS potentiometric map.

Other inputs included coefficients of transmissivity from the results of aquifer testing conducted at production wells in the vicinity of each WWTP and a storage coefficient of 0.0001, typical for a confined aquifer. Hydraulic gradients averaged around 0.0001 feet per year foot (ft/ft), with varying directions of groundwater flow based on the site's position within the USGS potentiometric map (see Figure 12.2).

A unique static groundwater elevation was entered for each WWTP site and modified slightly depending on whether injection wells were screened across two or three aquifer units of the PAS. As described by the assessment of local-specific capacity, injection wells were first simulated to screen the two adjoining aquifer units of the three aquifers (UPA and MPA; and MPA and LPA) comprising the PAS. This approach entailed adding the transmissivity of the two aquifers together and obtaining an average static water elevation for the two aquifers (USGS, 2016).

For wells screening the UPA and MPA, and the MPA and LPA, discrete simulations were conducted. As the MPA received effluent whether it was combined with the UPA or LPA, to effectively simulate well interference in the aquifer, each simulation involved all the wells. As an example, in simulations involving wells screening the UPA and MPA, wells screening the MPA, and LPA received one half of the total effluent flow.

To obtain the number of injection wells used in each simulation, maximum injection rates were held at 3 MGD per well and divided into the effluent flow HRSD projected for 2040. The injection wells were spaced (as much as possible) at roughly equal distance around the perimeter of the WWTP. Care was taken to avoid locating wells on existing structures. Locations were not evaluated for

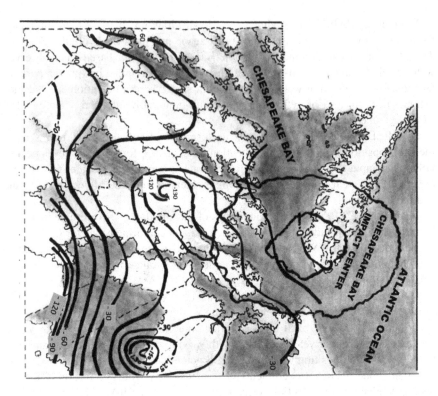

FIGURE 12.2 Potentiometric surface map of the Potomac Aquifer. Adapted from USGS 2015 in HRSD (2016). Illustration by F. R. Spellman and Kathern Welsh.

the practicality of positioning wells on lawns, parking lots, along fence lines, or other questionable areas that might host an injection well. Consistent with transient model runs conducted by USGS, CAPZONE simulations were set for a 50-year duration.

Simulated injection elevations were compared against the WWTP's threshold elevation in ft mean sea level, as defined by the ground surface elevation plus 10 psi (23 feet). Simulated injection levels from two aquifers exceeded the threshold elevation, which indicated that two aquifers could not facilitate effluent flows for the site. Accordingly, the simulation was run again combining all three aquifers of the PAS in each injection well at the WWTP.

Similar to the approach described previously, the transmissivities were added for each aquifer and static water-level elevations were averaged. In case the simulated injection levels continued to exceed threshold elevations, the effluent flow rates were reduced in each well until a solution was found where injection rates fell below the WWTP's designated threshold elevation. This approach resulted in determining the sustainable effluent rate for the site HRSD, 2016).

After the model runs that resolved the sustainable number of wells and effluent rates were completed, sensitivity testing was conducted to quantify the uncertainty in input parameters used in obtaining the model solutions. Sensitivity testing was conducted under conditions prevalent at the HRSD's York River Plant. Additional testing was conducted at the York River Plant to investigate the relationship between well spacing and interference. This testing proved particularly important to stations with elevated projected effluent flows, or smaller stations that could not support the number of wells to inject effluent at the projected 2040 rates. At these stations, HRSD could potentially locate injection wells at offsite locations at distances sufficient to lower the effects of hydraulic interference.

Modeling Results

After the model grid was set up, injection wells were located around the perimeter of each WWTP site, maximizing the number of wells given the site constraints. Through iterations of well layouts and injection rates the 2040 projected demands were tested at each WWTP. Table 10.5 shows the modeling results for each WWTP, including the maximum injection rates for sites that did not meet the 2030 or 2040 demand projections within the existing boundaries.

Army Base Treatment Plant

The Army Base WWTP model was able to meet the 2040 projected demands of 12 MGD using four wells. Two wells were screened in the UPA/MPA, with the other two set in the MPA/LPA, with all four-injecting effluent at 3 MGD. The injection head elevation reached a maximum of 9 ft mean sea level (MSL) within the UPA/MPA, falling several feet below the threshold elevation of 35 ft MSL. Depending on the aquifer used, the maximum draw-up found at the property boundary is approximately 61– 68 ft above static conditions.

Boat Harbor Treatment Plant

Seven injection wells spaced between 200 and 600 ft apart and screening all three PAS aquifers achieved an injection rate totally 14, falling short of the 16 MGD targeted 2040 flow projections. At 14 MGD, the injection level mgd is set at the maximum threshold value of 28 ft MSL. The total injection rate was constrained by the location and layout of the plant. The adjacent highway and the harbor limit the space available for wells, placing some wells at distances as close as 200 ft apart. If HRSD can find locations outside the WWTP boundaries, increasing well spacing to greater than 600 feet, six injection wells should prove sufficient to meet the 2040 flow projections. The maximum draw-up found at the property boundary fell approximately 93 ft above static conditions.

James River Treatment Plant

The James River WWTP model was able to meet the 2040 projected flows of 15 MGD using five wells. Unlike the Army Base model, the James River model required that the wells screened all three aquifers in the Potomac Aquifer System. Using the three PAS aquifers, the injection head elevation fell below the threshold elevation (42 ft MSL) by almost three feet. The maximum draw-up found at the property boundary totaled approximately 98 ft above static conditions.

Nansemond Treatment Plant

The 2040 projections for the Nansemond WWTP represented the second highest of any of the WWTP sites texted, reaching 28 MGD. The results of the model include using 12 well screening all three PAS aquifers to inject a maximum rate of 24 MGD but fell short of reaching the 2040 projections. At 24 MGD, the injection head elevation remained 2 ft below the threshold elevation (46.5 ft MSL). The large number of wells required and the limited space available led to the 4 MGD shortfall. The adjacent river and marsh limits the space available within the Nansemond WWTP site for locating injection wells. The maximum draw-up found at the property boundary was approximately 122 ft above static conditions.

Virginia Initiative Plant

The Virginia Initiative Plant's 2040 projections were the highest of all of the WWTP sites tested, requiring 33 MGD. The model included 14 UPA/MPA/LPA wells spread out across parcels north and east of the WWTP, comprising the golf course. The maximum attainable injection rate reached only 21 MGD, falling short of the 2040 projections. Meeting the projected 2040 flow of 33 MGD will require locating wells in offsite locations. The maximum projected draw-up found at the property boundary equaled approximately 101 ft above static conditions. In this scenario, draw-up was obtained from the boundary of the WWTP with the surrounding open space.

Williamsburg Treatment Plant

The Williamsburg Treatment Plant, like the Army Base Treatment Plant, was able to meet 2040 projections using five wells split between the UPA/MPA an MPA/LPA. For Williamsburg simulations, two wells were set in the UPA/MPA, with the other three in the MPA/LPA. Given the high threshold elevation at this site (82.5 ft MSL), the WWTP was able to exceed the 2040 demands (13 MGD).

At 15 MGD, injected through five wells, the injection elevation reached 53 ft MSL, well below the threshold elevation for the site. The maximum draw-up found at the property boundary totaled approximately 77–93 ft above static conditions, depending on the aquifer combination (UPA/MPA or MPA/LPA).

York River Treatment Plant

The York River Treatment Plant also successfully met the 2040 projections but required using all three PAS aquifers. The model was able to achieve 15 MGD (over 2040 projections of 14 MGD) using six UPA/MPA/LPA wells, reaching a simulated injection elevation of 29.5 ft MSL, matching the threshold value (29.5 ft MSL). The maximum draw-up found at the property boundary fell approximately 79 ft above static conditions.

SENSITIVITY OF AQUIFER PARAMETERS

Whenever mathematical models are used, as was the case in a phase of HRSD's SWIFT project, there is uncertainty in the inputs applied to get an output. Because of uncertainty in inputs and outputs used in mathematical modeling, a sensitivity analysis is called for; that is, it should be part of the entire process. With regard to the SWIFT project, a sensitivity analysis quantifies the doubt in a calibrated or predicted solution caused by uncertainty in the estimates of the aquifer parameters, injection stresses, and groundwater elevations. Basically, what a sensitivity analysis accomplishes is a process of recalculating outcomes under alternative assumptions and has a range of various purposes (Parnell, 1997), including:

- In the presence of uncertainty, it tests the strength of the results of the model.
- Amplified understanding of the relationships between input and output variables in a system to model.
- Further research can reduce uncertainty by identifying the model inputs that cause significant uncertainty in the output.
- Encountering unexpected relationships between inputs and outputs can be accomplished by searching for errors in the model.
- Model simplification.
- Increasing and enhancing the communication and the links from modelers and decision makers via using persuasion and straight talk.
- Employing Monte Carlo filtering to find regions in the space in input factors to optimum criterion.
- Knowing the sensitivity of parameters saves time by ignoring non-sensitive ones (Bahremand and Smedt, 2008).
- To develop better models, important connections between observations, model inputs, and predictions or forecasts must be identified (Hill et al., 2007; Hill and Tiedeman, 2007).

In HRSD's SWIFT program, the sensitivity of aquifer parameters was performed on the scenario that simulated injecting 14 MGD at the York River WWTP site. The York River site was chosen for selectivity analysis because it represents a site that accommodated the 2040 flows but required using all three PAS aquifers.

Additionally, changes in pumping stresses and groundwater elevations were tested on a single well, eliminating interference from multiple wells. Finally, two wells were simulated to measure interference at varying distances.

For this sensitivity analysis, characterizing the uncertainty of the modeled solution, input values for transmissivity, storativity, injection rate, simulation duration, and the groundwater elevations were systematically adjusted to assess how the changes affected groundwater elevations beneath the WWTP site. To quantify the evaluation, the maximum head generated from a sensitivity run was compared against the head from the original modeled solution.

The sensitivity analysis for aquifer transmissivity was performed by changing one parameter value at a time by −50 and +50% of the original parameter. The storage coefficient, injection rates, and groundwater elevations were tested by increasing and decreasing the values incrementally, not on a percent basis.

TRANSMISSIVITY

For purpose of clarity, in this text transmissivity is defined as the capacity of a rock to transmit water under pressure. The coefficient of transmissibility is the rate of flow of water, at the prevailing water temperature, in gallons per day, through a vertical strip of the aquifer one-foot wide, extending the full saturated height of the aquifer under a hydraulic gradient of 100%. A hydraulic gradient of 100-percent means a one-foot drop in head in one foot of flow distance.

With regard to HRSD's SWIFT project, the transmissivity applied to the combined units of the UPA, MPA, and LPA at York River Treatment Plant it was increased and decreased from the value used in the modeled solution. Values used in the sensitivity analysis ranged from 101,200 to 303,600 gpd/ft. The model was more sensitive to decreasing than increasing transmissivity by a factor approaching three times. Reducing transmissivity by 50% increased the maximum groundwater elevation at York River by 77 ft over the modeled solution of 29.5 ft MSL. Conversely, increasing the transmissivity value 50% decreased the mounding by only 27 ft from the modeled solution.

STORAGE COEFFICIENT

Storage coefficient is the volume of water released from storage in a unit prism of an aquifer when the head is lowered a unit distance. In the HRSD SWIFT model, the aquifer storage coefficient was increased and decreased from the value used in obtaining the modeled solution (0.0005). Storage coefficient used in the sensitivity analysis ranged from 0.00005 to 0.005. Reducing the storage coefficient by an order of magnitude increased the maximum head across the site by 13.6 ft. Conversely, increasing the storage coefficient by an order of magnitude decreased the maximum injection level by 13.5ft. As the modeled solution fell close to the threshold elevation, adjusting the storage coefficient can have significant effects on whether the UPA/MPA/LPA can except the 2040 injection rate.

INJECTION RATES

Sensitivity of changes to injection rates was tested by comparing the maximum injection levels of a single well at different injection rates. Injection rates within the sensitivity analysis ranged from 0 (static conditions) to 4 MGD. The model was almost identically sensitive to increases and decreases in injection rate. Reducing the rate from 3 MGD to 2 MGD reduced the maximum injection head by 8.3 ft while increasing the rate to 4 MGD increased the injection level by 8.2 ft.

SIMULATION DURATION

The duration of injection activity was adjusted to determine the model's sensitivity to changes of this parameter. An increase of 50% (75 years) over the original simulation (50 years) resulted in a rise in the maximum head value of 2.45 ft. Decreasing the simulation duration by 50% (25 years) lowered the maximum head value by 4.04 ft.

STATIC WATER LEVELS

The static water level was set at −56 ft MSL in the model solution. The model solution, with 14 MGD of injection into the UPA/MPA/LPA, reached a maximum groundwater elevation of 29.5 ft MSL. The static water level was raised and lowered 10 ft for the sensitivity analysis. The model appeared slightly more sensitive to an increase in the groundwater elevation than a decrease. Increasing the static water level to −46 ft MSL increased the maximum head at the well by 10.1 ft, decreasing the static water level to −66 ft MSL, reduced the maximum head by 9.9 ft.

WELL INTERFERENCE

To measure the effect of multiple wells injecting in proximity to each other, two wells were simulated with injection rates of 3 MGD under the subsurface conditions encountered at the York River WWTP for 50 years. The distance between wells was changed incrementally, while the maximum groundwater elevation was recorded at each well and at the midpoint between the wells. The well spacings tested ranged from 500 to 2,000 ft. Injection heads at each well ranged from −30.9 ft MSL with only Well 1 injecting to −12 ft MSL when spaced 500 ft apart. Between 500 and 3,000 ft of spacing, the maximum groundwater elevations in the wells varied about 4 ft. Interferences at Well 1 ranged from almost 19 ft to 15 ft, with Well 2 spaced from 500 to 3,000 ft away, respectively. At the midpoint between the wells, the head values ranged from −14.8 to −23.1 ft MSL, for distances of 500 to 3,000 ft, respectively. The interference in the PAS changed by 8.3 ft, or slightly less than 2 ft for every 500 ft of separation between the wells. The large amount of interference (greater than 15 ft) caused by a single nearby injection well even at relatively large spacings (3,000 ft) appears consistent with the results of the modeling at other WWTPs. WWTPs carrying large injection rats require many wells, each well increasing the injection levels at other wells and in the aquifer (HRSD, 2016).

HAMPTON ROADS REGION GROUNDWATER FLOW

The Virginia Coastal Plain Model (VCPM), a SEAWAT groundwater model, was employed to evaluate the hydraulic response of the PAS to injection operations at HRSD's seven WWTPs (Heywood and Pope, 2009; Langevin et al., 2008). This section presents the results of simulating injection at individual WWTPs and in scenarios with all seven WWTPs injecting simultaneously.

SEAWAT exemplifies a three-dimensional, variable density groundwater flow and transport model developed by the USGS based on MODFLOW and MT3DMS (Modular Three-Dimensional Multispecies Transport Model for Simulation). The VCPM groundwater model encompasses all the coastal plain within Virginia, and parts of the coastal plain in northern North Carolina and southern Maryland. The original VCPM was updated for use in the DEQ well permitting process and is now called VAHydro-GW. The VAHydro-GW model is discretized into 134 rows, 96 columns, and 60 layers. Most of the model cells are square with the horizontal edges measuring one mile. The upper 48 model layers are 35 ft thick each. Layer thicknesses for the lower model layers increase to 50 ft after layer 48 (top to bottom) and then 100 ft beneath layer 52.

The model simulates potentiometric water levels in 19 coastal plain hydrogeologic units. The water levels are simulated for each year from 1891 through 2012, based upon historic pumping records. The VAHydro-GW also simulates water levels for 50 years beyond 2012. These water levels are based upon two scenarios: the total permitted scenario and the reported use scenario. The total permitted scenario simulates water levels for 50 years beyond 2012 by using the May 2015 total permitted withdrawal rates established for withdrawal permits issued by the DEQ, together with the estimates for non-permitted (domestic wells, wells in Maryland and North Carolina, wells within unregulated portions of Virginia) withdrawals based upon 2012 estimated use. The total permitted scenario represents the estimated water levels 50 years into the future if all permittees within the coastal plain were to pump at their authorized maximum withdrawal rates for the duration of the 50-year period.

The reported use scenario simulates water levels for 50 years using pumping rates reported in 2012, for wells permitted by the DEQ and estimates for non-permitted withdrawals based upon 2012 estimated use. For most large permitted systems (greater than 1 MGD), reported pumping rates fall well below their total permitted diversion. The reported use simulation represents the best available estimate of water levels within the coastal plain aquifers over the next 50 years, if pumping were to continue at the currently reported pumping rates for the permitted wells within the coastal plain.

Virginia regulations have established limits on the amount of drawdown allowed as a result of permitted pumping within the coastal plain. The "critical surface" is defined as the surface that represents 80% of the distance between the land surface and the top of the aquifer. Individual model cells where simulated potentiometric water levels fall below the critical surface are referred to as "critical cells." Both the reported use and total permitted simulations show areas of the coastal plain for the Potomac, Virginia Beach, Aquia, Piney Point, and Yorktown-Eastover aquifers, where the predicted water levels at the end of the 50-year simulation end below the critical surface for those aquifers.

For any new or renewing permitted withdrawal, DEQ performs a technical evaluation which involves adding the proposed facility to the total permitted simulation. As a major criterion for permit issuance, the facility cannot create new critical cells in any aquifer due to their proposed withdrawal. The critical cells simulated at the end of the reported use simulation are not used for permit evaluation or issuance but represent a more plausible estimate of areas where water levels have lowered to crucial levels.

MODEL INJECTION RATES

VAHydro-GW row and column values were assigned to seven of HRSD's proposed injection WWTPs by using the well locations (latitude and longitude) to plot the position on a Geographic Information System coverage of the VAHydro-GW finite-difference grid. Each facility was simulated as a single point of injection and consequently assigned to only one row and one column. As explained earlier, each model cell is square with each cell edge measuring one mile. As a result, the rates injected through any number of wells are dependent on the individual WWTP. Because of course grid dimensions, the analysis is not intended to evaluate the number of injection wells that are required to dispose of injectate at each facility.

For the initial modeling, the well screen length for each WWTP was assumed to measure between 300 and 350 ft, thus screening across multiple layers of the VAHydro-GW model. Because the VAHydro-GW module utilizes the Hydrogeologic-Unit Flow (HUF) package, the model layers are independent of the hydrogeologic units. The HUF package is an alternative internal flow package that allows the vertical geometry of the system hydrogeology to be defined explicitly within the model using hydrogeologic units that can be different than the definition of the model layers. As for the model, a model layer may contain multiple hydrogeologic units. In order to ensure that simulated water levels were not artificially influenced by the Potomac confining unit, each injection well was assigned to the uppermost VAHydro-GW model layer filled by the PAS. The remainder of each injection well screen was assigned to lower, adjacent model layers.

MODELING DURATION

In addition to modeling each WWTP operating individually (except for the York River injection facility), all of the proposed facilities were modeled simultaneously at a combined flow of 114.01 MGD. The York River Treatment Plant was not included in the combined simulations because the facility lies within the outer rim of the Chesapeake Bay Bolide Impact Crater. As simulated in the VAHydro-GW, the horizontal hydraulic conductivity for the PAS at the cells within the bolide impact crater equal 0.0001 ft per day. As a result, simulated heads amounted to unrealistically high values when modeling injection at the York River Treatment Plant.

The individual WWTP scenarios and the combined WWTP scenario were simulated by adding the proposed injection rates to the total permitted and reported use simulations outlined previously, at the beginning of the 50-year predictive portion of those simulations (year 2013). The reported use and total permitted scenarios were also executed before adding the injection facilities to establish "baseline" conditions. This presentation refers to these scenarios as the "reported use baseline" and "total permitted baseline" simulations.

The model runs represent the DEQ's preferred metric for determining the beneficial impacts, if any, of proposed pumping/injection scenarios. The difference between water levels from an injection simulation and water levels from a baseline simulation represent the benefits, or recovery (rebound), resulting from the injection. The results of the injection and baseline simulations were compared at two points, 10 and 50 years into the predictive portion of the simulations.

ADVANCED WATER PURIFICATION

Hampton Roads Sanitation District is faced with a variety of future challenges related to the treatment and disposition of wastewater in its region of responsibility; this region includes much of the southern Chesapeake Bay, with its many major tributaries and surrounding communities. HRSD envisions that it can protect and enhance the region's groundwater supplies by reusing highly purified wastewater through advanced treatment and subsequent injection into the region's groundwater aquifers. For those of us who understand the natural water cycle, human-kind's urban water cycle, the proper use of various advanced wastewater treatment processes to purify the water, and HRSD's commitment to absolute excellence to its ratepayers and all those who live in the Hampton Roads region, as well as to restoring and sustaining the Chesapeake Bay—we understand that HRSD's vision not only has merit but is also necessary. It is necessary because the groundwater supply within the Potomac Aquifer is dwindling; it is in danger of contamination from saltwater intrusion, treated wastewater from HRSD's wastewater treatment plants that outfall nutrients into the Bay and contribute to dead zones and other environmental issues. Finally, HRSD understands that as native groundwater is withdrawn from the under-lying aquifers, this contributes to land subsidence in the region. Land subsidence plus global sea-level rise is the contributor to relative sea-level rise and if not abated will soon (in less than 150 years) inundate many of Hampton Roads major cities and other low-lying areas in the region.

In the previous chapter it was pointed out that HRSD and its contractor along with US Geological Survey (USGS) are working in unison to implement steps and model-ing to ensure the compatibility of treated injectate with native groundwater. It is also important to make sure the chemical match between injectate and native groundwater is safe for consumption. This is where advanced water treatment (AWT) comes into play. Wastewater treated only to conventional standards is probably safe enough for pipe-to-pipe connection but suffers from the public perception of the "Yuck Factor." That is, the old "your toilet water at my tap, no way, Jose and Maria," syndrome. HRSD's plans were published in the surrounding Hampton Roads area and the over-whelming majority of the populace voiced no objection to HRSD's plans. However, one of the few (but common) complaints was from those with wells drawing water from the Potomac Aquifer; they were worried that HRSD SWIFT project might contaminate their water source with toilet water. This is exactly what HRSD is work-ing hard to prevent by implementing advanced water treatment. Understand that the advanced water treatment process is in addition to normal wastewater treatment and filtration. In this chapter we describe the three treatment processes, reverse osmosis, nanofiltration, and biologically activated carbon, and the pilot studies that were used to determine which process (or processes) is best suited to facilitate HRSD's goal of producing and injecting the safest water possible.

BY THE BOOK, PLEASE!

The goal of most public service entities is to perform their functions by the book. The book in most cases is the written volume that contains the applicable regulations—the so-called laws of the land—that apply to their activities. For operations that can directly or indirectly affect the environment, the applicable regulations are generally federal-based and enforced (by the Environmental Protection Agency, for example). However, it is interesting to note that injection of reclaimed water into an aquifer that is used as a potable water supply is referred to as indirect potable reuse (IPR) and regulations have not been developed by the US Environmental Protection Agency (EPA) for potable reuse projects; therefore, states in which IPR is being practiced (or is being actively considered) have developed state-specific potable reuse regulations. About indirect potable water reuse compliance, it is the state regulator knocking at the door, not the Feds.

THOSE PLAYING BY THE BOOK IN INDIRECT POTABLE REUSE*

California and Florida have developed regulations governing the practice of indirect potable reuse (IPR). Other states allow IPR but establish project specific requirements on a case-by-case basis (e.g., Virginia, Texas). Because Virginia has not developed IPR regulations, but do allow IPR, the state will likely look to successful full-scale IPR projects within Virginia (e.g., Upper Occoquan Service Authority (USOA), Loudon Water) and other states' IPR regulations for guidance in regulating HJRSD's proposed direct injection IPR project.

ADDITIONAL DRINKING WATER CONSIDERATIONS

All IPR projects are required to comply with drinking water maximum contaminant levels (MCLs). This requirement is typically not difficult to meet because most modern wastewater treatment plants comply with drinking water MCLs (except for Tennessee, where nitrification and denitrification is not practiced). Contaminants of emerging concern (CEC) are defined as any synthetic or naturally occurring chemical or microorganism that is not commonly monitored in the environment but has the potential to enter the environment and cause known or suspected adverse ecological and/or human health effects. Personal Care Products and Pharmaceuticals (PCPPs) fall under this category. Pharmaceuticals and Personal Care Products comprise a very broad, diverse collection of thousands of chemical substances, including prescription and over-the-counter therapeutic drugs, fragrances, cosmetics, sunscreen agents, diagnostic agents, nutrapharmaceuticals, biopharmaceuticals, and many others. These emerging contaminants have garnered significant media attention in recent years because of improvements in analytical techniques allowing measurement of these chemicals at part per trillion levels (ppt). Although some impact on the ecology has been noted at a few WWTP discharge locations due to endocrine-disrupting compounds, no impact on human health has been observed. However, the best we can say about PCPPs and their effect on the environment and/or human health is that we do not know what we do not know about them.

Regarding the impact of endocrine disruptors, what we do know is that there is a growing body of evidence suggesting that humans and wildlife species have suffered adverse health effects after exposure to endocrine-disrupting chemicals (aka environmental endocrine disruptors). In this book, environmental endocrine disruptors

---

* Much of the material in this section is from HRSD (2016c). *Sustainable Water Recycling Initiative: Advanced Water Purification Process Feasibility Evaluation. Report 3.* Hampton Roads Sanitation District. Compiled by CH2M Newport News, VA.

are defined as exogenous agents that interfere with the production, release, transport, metabolism binding, action, or elimination of natural hormones in the body responsible for the maintenance of homeostasis and the regulation of developmental processes. This definition reflects a growing awareness that the issue of endocrine disruptors in the environment extends considerably beyond that of exogenous estrogens and includes anti-androgens and agents that act on other components of the endocrine system such as the thyroid and pituitary glands (Kavlock et al., 1996). Disrupting the endocrine system can occur in various ways. Some chemicals can mimic a natural hormone, fooling the body into over-responding to the stimulus (e.g., a growth hormone that results in increased muscle mass) or responding at inappropriate times (e.g., producing insulin when it is not needed). Other endocrine-disrupting chemicals can block the effects of a hormone from certain receptors. Still others can directly stimulate or inhibit the endocrine system, causing overproduction or underproduction of hormones. Certain drugs are used to intentionally cause some of these effects, such as birth control pills. In many situations involving environmental chemicals, an endocrine effect may not be desirable.

In recent years, some scientists have proposed that chemicals might inadvertently be disrupting the endocrine systems of humans and wildlife. Reported adverse effects include declines in populations, increases in cancers, and reduced reproductive function. To date, these health problems have been identified primarily in domestic or wildlife species with relatively high exposures to organo-chlorine compounds, including DDT and its metabolites, polychlorinated biphenyls (PCBs) and dioxides, or to naturally occurring plant estrogens (phytoestrogens). However, the relationship of human diseases of the endocrine system and exposure to environmental contaminants is poorly understood and scientifically controversial.

Although domestic and wildlife species have demonstrated adverse health consequences from exposure to environmental contaminants that interact with the endocrine system, it is not known if similar affects are occurring in the general human population, but again, there is evidence of adverse effects in populations with relatively high exposures. Several reports of declines in the quality and decrease in the quantity of sperm production in humans over the last five decades and the reported increase in incidences of certain cancers (breast, prostate, testicular) that may have an endocrine-related basis have led to speculation about environmental etiologies (Kavlock et al., 1996). For example, Carlson et al. (1992) point to the increasing concern about the impact of the environment on public health, including reproductive ability. They also point out that controversy has arisen from some reviews which have claimed that the quality of human semen has declined. However, only little notice has been paid to these warnings, possibly because the suggestions were based on data on selected groups of men recruited from infertility clinics, from among semen donors, or from candidates for vasectomy. Furthermore, the sampling of publications used for review is not systematic, thus implying a risk of bias. As a decline in semen quality may have serious implications for human reproductive health, it is of great importance to elucidate whether the reported decrease in sperm count reflects a biological phenomenon or, rather, is due to methodological errors.

Data on semen quality collected systematically from reports published world-wide indicate clearly that sperm density had declined appreciably from 1938 to1990, although we cannot conclude whether this decline is continuing. Concomitantly, the incidence of some genitourinary abnormalities including testicular cancer and possibly also maldescent (faulty descent of the testicle into the scrotum) and hypospadias (abnormally placed urinary meatus) (opening in male penis) have increased. Such remarkable changes in semen quality and the occurrence of genitourinary abnormalities over a relatively short period is more probably due to environmental rather than genetic factors. Some common prenatal influences could be responsible both for the decline in sperm density and for the increase in cancer of the testis, hypospadias, and cryptorchidism (one or both testicles fail to move to scrotum). Whether estrogens or compounds with estrogen-like activity or other environmental or endogenous factors damage testicular function remains to be determined (Carlson et al., 1992). Even though we stated that we do not know what we do not know about endocrine disruptors, it is known that the normal functions of all organ systems are regulated by endocrine factors, and small disturbances in endocrine function, especially during certain stages of the life cycle such as development, pregnancy, and lactation, can lead to profound and lasting effects. The critical issue is whether sufficiently high levels of endocrine-disrupting chemicals exist in the ambient environment to exert adverse health effects on the general population.

Current methodologies for assessing, measuring, and demonstrating human and wildlife health effects (e.g., the generation of data in accordance with testing guideline) are in their infancy. USEPA has developed testing guidelines and the Endocrine Disruption Screening Program (EDSP; discussed later in this text), which is mandated to use validated methods for screening the testing chemicals to identify potential endocrine disruptors, determine adverse effects, dose-response, assess risk, and ultimately manage risk under current laws. The best way in which to end this brief discussion is to provide a statement by someone who really knows and understands endocrine disruptors and the potential impact on humans, wildlife, and the environment in general.

"Large numbers and large quantities of endocrine-disrupting chemicals have been released in the environment since World War II. Many of these chemicals can disturb development of the endocrine system and of the organs that respond. to endocrine signals in organisms indirectly exposed during prenatal and/or early postnatal life; effects of exposure during development are permanent and irreversible."

—Theo Colburn et al., 1993

Regardless of our current lack of definitive evidence on the impact of PCPPs on human health, multiple barriers, and relatively low total organic carbon (TOC) limits (0.5 mg/L-3 mg/L) have been established for most IPR projects to limit the presence and concentration of CECs. In addition, some states require specific treatment (e.g., advanced oxidation) to ensure oxidation of a large portion of CECs.

Other nonregulated water quality parameters are often considered when implementing an IPR project, including the following:

- **Nitrosamines:** Nitrosamines are chemical compounds of the chemical structure $R^1N(-- R^2)--N=O$; they are suspected carcinogens and have recently been showed to form during wastewater treatment, primarily through the disinfection process when chloramines react with n-nitrosodimethylamine (NDMA) precursors such as secondary, tertiary, and quaternary amines present in the water. California has established a notification level for NDMA of 10 nanograms per liter (ng/L), and USEPA is considering regulating some of the nitrosamines under the Safety drinking Water Act as part of the Contaminant Candidate List 3 process, which may result in limitations applied to IPR projects.
- **Total Hardness:** Total hardness in WWTP secondary effluent can often exceed levels generally deemed aesthetically acceptable by the public unless treatment or blending is specifically provided.

Note: Compatibility between the aquifer and the injected water is also required to avoid scale formation that may plug injection wells or release undesirable compounds from the soil matrix (e.g., arsenic).

ADVANCED WATER TREATMENT PROCESSES

Advanced treatment provided at indirect potable reuse plants varies but is typically focused on providing multiple barriers for the removal of pathogens and organics. Nitrogen and total dissolved solids (TDS) removal is provided at some locations where necessary. Water extracted from direct injection and surface spreading projects that recharge groundwater isn't typically treated again prior to distribution into the potable water system; however, water from surface augmentation projects is typically treated again at water treatment plants because of water treatment requirements stipulated by USEPA's Surface Water Treatment Rule. For example, Fairfax County's Griffith Water Treatment Plant provides coagulation, sedimentation, ozone oxidation, biologically activated carbon filtration, and chlorine disinfection for water extracted from the Occoquan Reservoir that is augmented by Upper Occoquan Service Authority's (UOSA's) indirect potable reuse plant.

Treatment provided for indirect potable reuse projects is typically a combination of multiple barriers for the removal of pathogens and organics. Multiple barriers for pathogens are typically provided through a combination of coagulation, flocculation, sediment, lime clarification, filtration (granular or membrane), and disinfection (chlorine, ultraviolet [UV], or ozone). Multiple barriers for organics removal are typically provided through a combination of advanced treatment processes (e.g., reverse osmosis, granular activated carbon [GAC], ozone in combination with GAC [biologically activated carbon]), although conventional treatment processes (e.g., coagulation, softening) also provide removal at some locations. All potable reuse plants and/or processes discussed in this presentation include a robust organics removal process of either GAC, reverse osmosis, or either GAC, reverse osmosis, or soil aquifer treatment (SAT), which are effective barriers to bulk and trace organics, and represent the backbone of the potable treatment process. SAT land treatment is the controlled application of wastewater to earthen basins in permeable soils at a rate typically measured in terms of meters of liquid per week. The purpose of a soil aquifer treatment system is to provide a receiver aquifer capable of accepting liquid intended to recharge shallow groundwater. System design and operating criteria are developed to achieve that goal. However, there are several alternatives with respect to the utilization or final fate of the treated water (USEPA, 2006):

- Groundwater recharge.
- Recovery of treated water for subsequent reuse or discharge.
- Recharge of adjacent surface streams.
- Seasonal storage of treated water beneath the site with seasonal recover for agriculture.

The SAT process typically includes application of the reclaimed water using spreading basins and subsequent percolation through the vadose zone. SAT provides significant removal of both pathogens and organics through biological activity and natural filtration. However, because the Potomac Aquifer is confined, providing SAT for treatment through the vadose zone to recharge the aquifer is not possible for the HRSD SWIFT project. On the other hand, movement of reclaimed water through the aquifer after direct injection will provide significant treatment benefits, including excellent removal of pathogens, reverse osmosis-, and granular activated carbon. Advanced water treatment plants are often provided at locations where SAT treatment through the vadose zone is not feasible because these processes can be implemented at most locations.

Reverse osmosis- and GAC-based advanced treatment trains were developed for the HRSD groundwater recharge project using the historical WWTP effluent water quality data and the preliminary aquifer recharge water quality goals discussed previously. Three treatment trains were developed from this analysis and include a RO-based train, a nanofiltration (NF)-based train, and a GAC-based train (Figure 12.3). Consideration for each of these treatment trains includes the following (HRSD, 2016c):

- **RO-Based Train:** RO has become common for potable water reuse projects in California and many international locations (e.g., Singapore, Australia) because of its effective removal of TDS, TOC, and trace organics. California regulations require the use of RO for direct injection reuse projects or a comparable alternative with regulatory approval. RO creates a waste (concentrate) stream that can be difficult and costly to dispose of, especially at inland locations. Most locations where RO has been implemented are located near the ocean where disposal of RO concentrate is convenient and much less costly than at inland locations.

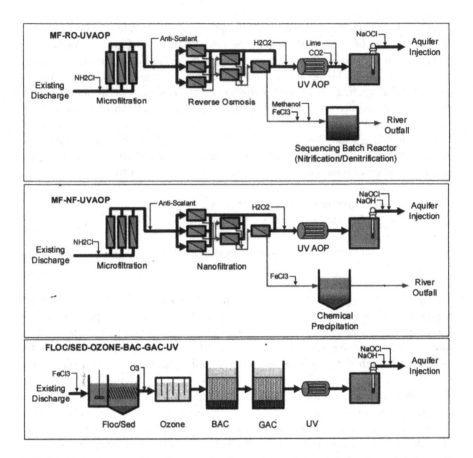

FIGURE 12.3 Process flow diagrams for three advanced wastewater treatment processes. Source: HRSD/CH2M (2016). CH2M Report No. 3.

- **NF-Based Train:** The NF-based train is like the RO-based train but operates at significantly lower pressure and generates a less saline concentrate, which results in significant cost savings. This process does not meet California's IPR regulatory requirements but provides excellent treatment with significant removal of pathogens and organics. This train provides partial TDS removal by providing a high level of removal of divalent ions (e.g., calcium, magnesium) and moderate removal of monovalent ions (e.g., sodium chloride). NH_3 and NOx-N removal is much lower with NF compared to RO, which results in lower total nitrogen (TN) concentration in the concentrate.

- **GAC-Based Train:** This train is a modernized version of full-scale operational IPR plants that have successfully been in operation for decades in Virginia (1978, 2008), Texas (1985), and more recently, Georgia (2000). GAC adsorption is used as the backbone process for organics removal, and other treatment has been added for multiple barriers to pathogens and organics. Flocculation and sedimentation provide removal of solids, pathogens, organics, and phosphorous. Ozone provides disinfection of pathogens and oxidation of organics, including oxidation of CECs and high molecular weight organic matter to smaller organic fractions that can be assimilated by biological activity present on GAC media, which is referred to as biologically active carbon (BAC) filtration. This treatment train does not provide any TDS removal and, therefore, does not generate a TDS-enriched waste that might require further treatment prior to discharge.

TREATMENT PLANT EFFLUENT WATER QUALITY

HRSD provided historic effluent water quality data for seven WWTPs to identify specific water quality challenges requiring treatment. The WWTPs analyzed include Army Base, Boat Harbor, James River, Nansemond, VIP, and York River.

DATA SOURCES FOR EVALUATION

The following three primary data sources were used in the evaluation:

- **2013/2014 Water Quality Data:** Detailed water quality data were provided for each WWTP effluent for October 2013 through September 2014. The data were provided as raw data in tables. With few exceptions, these data included the following parameters at the state frequency: chloride (1x/week), calcium (2x/week), magnesium (2x/week), potassium (2x/week), sodium (2x/week), total alkalinity (3x/week), 5-day biochemical oxygen demand (BOD$_5$; 5x/week), pH (1x/day), turbidity (1x/day), ammonia-nitrogen (NH$_3$-N; 2x/week), NOx-N (4x/week), Orthophosphate (3x/week), TKN (4x/week), total phosphorus (TP; 5x/week), and TSS (5x/week).
- **2011, 2012, and 2013 Water Quality Data:** Detailed influent and effluent water quality data were provided for each WWTP from the 2011, 2012, and 2013 HRSD Wastewater Characteristics Studies (no James River data were provided for 2013). The data provided were presented as minimum, average, and maximum values. The data included flow (continuous), temperature (3x/day), pH (12x/day), total alkalinity (1x/week), BOD$_5$ (5x/week), TSS (5x/week), turbidity (5x/week), fecal coliform (5x/week), TKN (frequency not reported), NOx-N (frequency not reported), and TP (frequency not reported). Data were also provided for influent chloride (1x/week) and influent sulfate (4x/week), selected heavy metals (1x/year), and a variety of organics (volatile, base/acid, pesticides; total trihalomethanes [TTHMs] 1x/year).
- **2014 Total Dissolved Solids Data:** Effluent TDS data were provided for each WWTP for January through September 2014. The data were provided as raw data in tables, and data points were provided once per week.

Effluent chemical oxygen demand (COD) and TOC data collected by HRSD on a weekly basis from February 2015 through April 2015 were also used in the evaluation.

Data Evaluation

Evaluation of effluent quality from each of the seven treatment plants involved in the HRSD's SWIFT project included identification of the strength of each data source and was qualitatively documented as excellent, good, and limited. Excellent data included detailed 2013/2014 raw data and minimum/average/maximum annual data from 2011 through 2013. Good data included a full data set from only one of the sources. Limited data included data that were only collected once per year.

- **Total Dissolved Solids:** The drinking water secondary MCL for TDS is 500 mg/L. The average effluent TDS from each WWTP except James River exceeds 500 mg/L; Army Base (1,292 mg/L) and VIP (853 mg/L) have notably high TDS concentrations.
- **Ammonia:** Army Base and Boat Harbor plants have an average effluent NH_3 of 25.2 and 16.0 mg N/L, respectively, although the Army Base plant was recently upgraded to biological nutrient removal (BNR) and now produces effluent with low ammonia and total nitrogen concentrations that are comparable to HRSD's other BNR plants. TN concentrations in excess of 10 mg/L are typically not allowed for groundwater recharge into potable aquifers; therefore, additional nitrogen treatment would likely be required at the Boat Harbor WWTP.
- **Total Trihalomethanes:** Elevated TTHM levels were recorded at Nansemond (82.4 micrograms per liter [μg/L]) and Williamsburg (64.7 μg/L). TTHM levels at the other plants ranged from 3 μg/L to 50 μg/L. More TTHM data should be collected as the data sources used were limited. The drinking water Primary MCL for TTHMs is 80 μg/L.
- **Total Hardness:** Hardness in water is caused by the presence of certain positively charged metallic ions in solution in the water. The most common of these hardness-causing ions are calcium and magnesium; others include iron, strontium, and barium. The two primary constituents of water that determine the hardness of water are calcium and magnesium. If the concentration of these elements in the water is known, the total hardness of the water can be calculated. To make this calculation, the equivalent weights of calcium, magnesium, and calcium carbonate must be known; the equivalent weights are given below.

EQUIVALENT WEIGHTS

Calcium, Ca	**20.04**
Magnesium, Mg	**12.15**
Calcium Carbonate CaCO₃	**50.045**

Calcium and magnesium ions are the two constituents that are the primary cause of hardness in water. To find total hardness, we simply add the concentrations of calcium and magnesium ions, expressed in terms of calcium carbonate, $CaCO_3$, using Equation (12.3).

$$Total\,Hardness, mg/L\,as\,CaCO_3 = Cal.Hard., mg/L\,as\,CaCO_3$$
$$+Mg.\,Hardness., mg/L\,as\,CaCO_3 \tag{12.3}$$

As mentioned, total hardness is comprised of calcium and magnesium hardness. Once total hardness has been calculated, it is sometimes used to determine another expression of hardness—carbonate and noncarbonate. When hardness is numerically greater than the sum of bicarbonate and carbonate alkalinity, that amount of hardness equivalent to the total alkalinity (both in units of mg $CaCO_3/L$) is called the *carbonate hardness*; the amount of hardness in excess of this is the *noncarbonate hardness*. When the hardness is numerically equal to or less than the sum of carbonate and noncarbonate alkalinity, all hardness is carbonate hardness and noncarbonate hardness is absent. Again, the total hardness is comprised or carbonate hardness and noncarbonate hardness:

$$Total\,Hardness = Carbonate\,Hardness + Noncarbonate\,Hardness \tag{12.4}$$

During the evaluation, total hardness data were not specifically provided, but it was calculated using the detailed calcium and magnesium data. Total hardness in drinking water systems is often limited to 150 mg/L (as $CaCO_3$) or less to avoid customer complaints; and Boat Harbor (161 mg/L), VIP (181 mg/L), and York River (194 mg/L) all show average effluent data above this value. Total hardness concentrations in potable water in the surrounding area are less than the secondary MCL of 500 mg/L.

- **Dissolved Organic Carbon and Soluble Chemical Oxygen Demand:** DOC and sCOD are important parameters to measure for potable reuse plants because advanced treatment goals and regulatory requirements are often developed for these constituents. The average DOC and sCOD concentrations in the effluent from the seven WWTPs range from 8.6 to 11.2 mg/L and 25 to 49 mg/L, respectively, which are within the typical range for WWTPs practicing biological nutrient removal. TOC concentrations above 10 mg/L become increasingly difficult to treat tor recommended levels for certain advanced treatment trains so additional DOC sampling and bench and pilot testing is recommended to confirm adequate treatment performance.
- **Heavy Metals, VOCs, Synthetic Organic Chemicals, and Other Organics:** The 2011, 2012, and 2013 data set included minimum, average, and maximum data for heavy metals, volatile organic chemicals, synthetic organic chemicals, and other organics. Although the data sets were limited (typical frequency of once per year), the data were compared to the applicable drinking water MCLs in order to identify any potential contaminants of concern. The recorded data that were compared to drinking water MCLs

include antimony, arsenic, beryllium, cadmium, chromium, copper, lead, mercury, selenium, thallium, benzene, TTHMs, carbon tetrachloride, chlorobenzene, 1,2-dichloroethane, 1,1-dichloroethylene, 1,2-dichloropropne, ethylbenzene, toluene, 1,1,1-trichlorotehtan, 1,2,2-trichloroethand, trichloroethane, and vinyl chloride.

- There are additional parameters with regulated MCLs that did not have any data provided for these WWTPs. Consequently, additional regular sampling for all MCLs, EPA Candidate Contaminant List 3 parameters, and other chemicals of concern at each WWTP is recommended. In addition, a comprehensive study identifying industrial and commercial facilities that discharge to the wastewater collection system is advisable to identify potential locations where chemical contamination could occur that could negatively affect finished water quality or treatment processes if not already known. An outreach program aimed at limiting discharge of contaminants of concern and/or additional water quality monitoring at specific locations may need to be implemented if aquifer replenishment is ultimately pursued and implemented.

In addition to the average WWTP effluent data collected, selected 99th percentile effluent data from the 2013/2014 data set were analyzed to determine peak loadings from the WWTP that could be problematic for various treatment processes. Peak loadings can either be accounted for by selected I treatment process that is designed for the maximum values or by providing a large enough equalization volume of primary or secondary effluent to attenuate the loading. The following selected parameters of concern based on the 99th percentile data which include the following:

- **Nitrate/Nitrite-Nitrogen:** Average effluent NOx-N levels were well under the nitrate MCL of 10 mg-N/L; however, 99th percentile data at VIP (10.5 mg/L) and Williamsburg (.3 mg/L) show that NOx-N levels could periodically exceed the nitrate MCL, which could require NOx-N specific treatment or additional storage. The 99th percentile NOx-N concentration at Boat Harbor is also high, but BNR is not currently practicing at this plant. When BNR is implemented at Boat Harbor, the variability of the NOx-N data should be re-evaluated.
- **Biochemical Oxygen Demand and Total Suspended Solids:** High 99th percentile BOD_5 and TSS levels suggest occasional plant upsets that could be problematic for filtration (granular or membrane). This could require increased storage or treatment or automated monitoring to divert flow away from the AWTP during high biochemical oxygen demand (BOD) and TSS loadings.
- **Total Dissolved Solids:** WWTP effluent TDS values are not expected to fluctuate significantly, yet Army Base, VIP, and Boat Harbor each show 99th percentile values that are significantly higher than the average. Periodically high TDS values could violate treatment goals if reverse osmosis is not selected and would require additional storage or provision for divisions.

ADVANCED TREATMENT PRODUCT WATER QUALITY

INORGANIC WATER QUALITY

Using the historical water quality data presented previously and the expected performance of each unit process based on professional judgment, mass balance calculations for key inorganic parameters were performed for each treatment train at seven of HRSD's WWTPs. Summary tables for each treatment train are provided in Tables 12.3, 12.4, and 12.5. Detailed mass balances reveal the following:

- The RO-based treatment process provides the lowest concentration of all water quality parameters. However, treatment to this level may not be necessary in all cases. For example, the finished water TDS concentration is about 50 mg/L, which is well below the secondary MCL (500 mg/L) and the minimum background TDS in the Potomac aquifer (~750 mg/L). The very low TDS reverse osmosis permeate may increase mobilization of trace metals in the aquifer, which is undesirable.
- The NF-based treatment process provides excellent water quality as shown in Table 7.3. The NF process removes very little nitrogen; therefore, the TN concentration in the finished water exceeds the recommended upper range (10 mg/L) at Boat Harbor and is approaching the 10-mg/L limit at two other WWTPs (VIP and Williamsburg). Nitrification and denitrification improvements at these WWTPs may be necessary to ensure regular compliance with the recommended TN limit. Alternatively, NF membranes that have higher nitrogen removal can be considered; however, their use will result in a higher TDS concentrate stream.
- The GAC-based treatment process provides excellent water quality as shown in Table 12.5. Specific considerations related to this process include the following:
- Although some incidental nitrification may occur in the biologically activated carbon filters, the process is not intended, nor typically designed, to remove nitrogen. Therefore, nitrogen removal should be considered at the upstream WWTPs which will require nitrification and denitrification improvements at several plants to ensure regular compliance with the recommended TN limit.
- No TDS are removed through this process. The Army Base and VIP plants have elevated TDS that regularly exceed 750 mg/L. Upstream mitigation, such as reducing infiltration and inflow in areas with high TDS or eliminated industrial discharge high in TDS, may be required at these locations if a TDS limit of 750 mg/L is established.
- Hardness removal with chemical precipitation may be required at three plants (Boat Harbor, VIP, and York River), although more investigation is necessary to determine if the total hardness at these plants is acceptable (161–194 mg/L as $CaCO_3$) from aesthetic and aquifer geochemistry perspectives. If not, the proposed flocculation-sedimentation process shown for the GAC-based treatment train could be modified to a chemical softening process for those plants with elevated hardness.

TABLE 12.3
Projected Average Finished Water from the RO-Based Treatment Train—Inorganics

Parameter	Unit	Recommended Range	AB	BH	JR	NP	VIP	WB	YR
Avg. Flow	mgd	N/A	9.26	11.90	11.30	14.02	25.49	7.56	10.28
Alkalinity	mg/L	40 - 150	40	40	40	40	40	40	40
TDS	mg/L	0 - 750	87	46	37	52	61	47	46
Hardness mg/L	CaCO$_3$	50 - 150	43	43	42	43	43	42	44
TN	mg/L	0 - 10	4.84	3.95	0.91	1.01	1.31	1.27	0.89
NOx-N	mg/L	0 - 8	0.02	0.06	0.47	0.57	1.02	0.81	0.48
TP	mg/L	0 – 1	0.005	0.0035	0.004	0.01	0.00	0.01	0.00

Source: CH2M/HRSD (2016).

TABLE 12.4

Projected Average Finished Water from the NF-Based Treatment Train—Inorganics

Parameter	Unit	Recommended Range	AB	BH	JR	NP	VIP	WB	YR
Avg. Flow	mgd	N/A	9.26	11.90	11.30	14.02	25.49	7.56	10.28
Alkalinity	mg/L	40 - 150	67	77	45	85	31	87	57
TDS	mg/L	0 - 750	642	280	191	333	415	283	262
Hardness mg/L	CaCO$_3$	50 - 150	65	76	68	57	68	67	47
TN	mg/L	0 - 10	<10	15.7	4.8	5.5	7.9	7.2	48
NOx-N	mg/L	0 - 8	0.10	0.40	3.1	3.8	6.7	5.3	3.2
TP	mg/L	0 - 1	0.05	0.04	0.04	0.06	0.03	0.06	0.03

Source: CH2M/HRSD (2016).

Note: Highlighted cells indicated areas of concern; modifications to WWTP operations may be required and/or additions treatment at the AWTP.

TABLE 12.5

Projected Average finished Water from the GAC-Based Treatment Train—Inorganics

Parameter	Unit	Recommended Range	AB	BH	JR	NP	VIP	WB	YR
Avg. Flow	mgd	N/A	10.90	14.00	13.30	16.50	30.00	8.89	12.09
Alkalinity	mg/L	40–150	80	99	38	132	13	118	79
TDS	mg/L	0–750	1,422	616	420	734	918	623	615
Hardness	mg/L CaCO$_3$	50–150	143	161	99	66	181	97	194
TN	mg/L	0–10	<10	23	5.7	6.4	8.5	8.1	5.6
NOx-N	mg/L	0–8	0.10	0.40	3.1	3.8	<5	5.4	3.2
TP	mg/L	0–1	0.50	0.50	0.50	0.50	0.50	0.50	0.50

Source: CH2M/HRSD (2016).

Note: Highlighted cells indicate areas of concern; modifications to WWTP operations may be required and/or additions treatment at the AWTP.

ORGANIC WATER QUALITY

Bulk Organics

The application of robust treatment barriers for the removal of organics has historically been a center tenant in the implementation of full-scale potable reuse projects. These treatment barriers address the presence of unknown organic compounds of chronic health concern that may be present in the secondary effluent—a significant and pressing part of the old "we do not know what we do not know" syndrome. Regulations and permits for potable reuse projects have been developed by establishing limits on bulk organic parameters, such as COD and TOC, which act as surrogates for organic compounds of wastewater origin. Virginia established a COD limit of 10 mg/L for the Occoquan and Dulles areas Watershed Polices, which apply to the UOSA IPR project (constructed in 1978) and the Broad Run WRF project (constructed in 2008), respectively. California and Florida have established TOC limits of 0.5 mg/L and 3 mg/L, respectively, in their IPR regulations.

The advanced water treatment plant's finished water COD and TOC concentrations that would need to comply with the established permit limit is dependent on the initial concentration in the WWTP effluent and the specific treatment processes employed at the advanced water treatment plant. Table 12.6 shows the estimated finished water TOC concentration from each of the three proposed advanced water treatment plant treatment trains (i.e., RO, NA, and BAC) when treating effluent from each of HRSD's WWTPs. The calculations use full-scale advanced water treatment plant effluent TOC and DOC sampling and treatment process pilot testing. The following can be concluded from the information in the table:

- Compliance with a California-based TOC limit of 0.5 mg/L could only be achieved by implementing an RO-based treatment train.
- Compliance with a Florida-based UOSA-type permit (3 mg/L TOC and 10 mg/L COD, respectively) could likely be achieved at most WWTPs by

TABLE 12.6
Estimated TOC Concentration in WWTP Effluent and AWTP Finished Water

Location	AB	BH	JR	NP	VIP	WB	YR
WWTP Effluent DOC	9.9	9.8	9.0	11.2	8.6	10.1	9.5
RO-AWTP	0.3	0.3	0.3	0.3	0.3	0.3	0.3
NF-AWTP	1.0	1.0	0.9	1.1	0.9	1.0	1.0
GAC-AWTP	2.5	2.5	2.3	2.8	2.2	2.6	2.4

Source: HRSD/CH2M (2016).
Notes:
Preliminary TOC Goa: 0.5–3 mg/L. The following DOC removal percentages were used in the table based on full-scale treatment performance data: 97% for RO; 9-0% for NF; 35% for flocculation + sedimentation; 40% for ozone + BAC; and 40% for GAC adsorption

any of the three proposed treatment trains, and by a hybrid treatment train that combined partial RO treatment with GAC-based treatment.

- The GAC-based AWTP will require regular replacement for regeneration of the GAC media to provide consistent TOC removal. Pilot testing is necessary to determine the GAC regeneration frequency requirements.
- Measurement of the TOC and DOC in the final effluent from each WWTP is recommended on a regular basis to accurately determine TOC removal requirements.

The following DOC removal percentages were used in this based on full-scale treatment performance data: 97% for RO; 90% for NF; 35% for flocculation + sedimentation; 40% for ozone + BAC; and 40% for GAC adsorption.

Trace Organics

Earlier, PCPPs were mentioned along with other CECs. Additional concerns about CECs continue to be raised about the potential for, yet generally unknown, chronic human health effects related to the thousands of organic chemicals that may end up in wastewater effluent at trace levels (mg/L). Furthermore, the efficacy of conventional water treatment processes that may end up treating source waters that have some effluent contribution is typically low. Each advanced treatment process considered in this discussion differs in its effectiveness at removing CECs. Research has shown that RO-based, NF-based, and GAC-based potable reuse treatment trains provide multiple unit processes that are effective barriers to a wide range of CECs. The RO- and NF-based treatment trains provide substantial removal through membranes (RO/NF) and advanced oxidation (UVAOP), while the GAC-based treatment train provides significant removal through ozone-biologically active granular activated carbon (BAC) and GAC. Representative removals by advanced treatment processes for a variety of CECs was determined through recent research and monitoring of full-scale treatment facilities. These processes are redundant in the removal of some CECs (provide multiple barriers to their passage) and are complementary in the removal of others. For example, both ozone and GAC are effective barriers to the anticonvulsant drug carbamazepine, but only GAC acts as an effective barrier to the flame retardant tris (2-cloroethyll) phosphate (TCEP). No one process provides complete removal of all compounds, but RO generally provides the best removal of a wide range of compounds. However, these compounds are not destroyed or transformed by RO, but transferred to the RO concentrate (at high concentration); and thus, their presence in the concentrate must be considered, particularly where the concentrate is discharged to a receiving water body.

At the present time, treatment for all CECs does not appear to be a differentiator among potable reuse treatment trains. Although health effects of many CECs—either alone or as a mixture—have not been demonstrated at the ng/L concentrations typically detected in wastewater effluent, the proposed treatment trains do reduce the concentrations of many of these chemicals to a significant degree. Meanwhile, USEPA is prioritizing and studying several chemicals through their candidate contaminant list program.

RO Concentrate Disposal*

To this point in the book, we have discussed the benefits of reverse osmosis operating systems. It is important to point out, however, that along with the good, there is the not so good; that is, RO systems have their advantages, but they also have a few disadvantages. The one disadvantage pointed out and discussed here is the major one; that is, concentrate disposal. Where is the concentrate waste stream to be disposed of?

Mass Balance

To gain better understanding of RO concentrate disposal issues and techniques we begin with a discussion of mass balance. The simplest way to express the fundamental engineering principle of *mass balance* is to say, "Everything has to go somewhere." More precisely, the *law of conservation of mass* says that when chemical reactions take place, matter is neither created nor destroyed. What this important concept allows us to do is track materials (concentrates), that is, pollutants, microorganisms, chemicals, and other materials from one place to another. The concept of mass balance plays an important role in reverse osmosis system operations (especially in desalination) where we assume a balance exists between the material entering and leaving the RO system: "what comes in must equal what goes out." The concept is very helpful in evaluating biological systems, sampling and testing procedures, and many other unit processes within any treatment or processing system.

All desalination processes have two outgoing process streams—the product water which is lower in salt than the feed water, and a concentrated stream that contains the salts removed from the product water. Even distillation has a "bottoms" solution that contains salt from the vaporized water. The higher concentrated stream is called the "concentrate." The nature of the concentrate stream depends on the salinity of the feed water, the amount of product water recovered, and the purity of the product water. To determine the volume and concentration of the two outgoing streams, a mass balance is constructed. The recovery rate of water, the rejection rate of salt, and the input flow and concentration are needed to solve equations for the flow and concentration of the product and concentrate.

RO Concentrate Disposal Practices†

RO concentrate is disposed of by several methods, including surface water discharge, sewer discharge, deep well injection, evaporation ponds, spray irrigation, and zero liquid discharge.

Surface Water and Sewer Disposal   Disposal of concentrate to surface water and sewer are the two most widely used disposal options for both desalting membrane

---

* Adapted from F.R. Spellman (2015). *Reverse Osmosis: A Guide for the Non-Engineering Professional.* Boca Raton, FL: CRC Press.

† Material in this section is from US Department of Interior Bureau of Reclamation (Mickley & Associates) (2006) *Membrane Concentrate Disposal: Practices and Regulation.* Washington, DC.

processes. Data from the present survey (post 1992 data only) provide the following statistics:

Disposal Option	Desalting Plants (%)
Surface Water Disposal	45
Disposal to Sewer	42
Total	87

This disposal option, though not always available, is the simplest option in terms of equipment involved and is frequently the lowest cost option. As will be seen, however, the design of an outfall structure for surface water disposal can be complex.

Disposal to surface water involves conveyance of the concentrate or backwash to the site of disposal and an outfall structure that typically involves a diffuser and outlet ports, or valve mounted on the diffuser pipe. Factors involved in the outfall design are discussed in this section, and cost factors are presented. However, due to the large number of cost factors and the large variability in design conditions associated with surface water disposal, a relatively simple cost model cannot be developed. Disposal to surface waters requires a National Pollutant Discharge Elimination System (NPDES) permit.

Disposal to the sewer involves conveyance to the sewer site and typically a negotiated fee to be paid to the WWTP. Because the negotiated fees can range from zero to substantial, there is no model that can be presented. No disposal permits are required for this disposal option. Disposal of concentrate or backwash to the sewer, however, affects the WWTP's effluent that requires an NPDES permit.

About design considerations for disposal to surface water, a brief discussion of ambient conditions, discharge conditions, regulations, and the outfall structure are discussed herein.

Because receiving waters can include rivers, lakes, estuaries, canals, oceans, and other bodies of water, the range of ambient conditions can vary greatly. *Ambient conditions* include the geometry of the receiving water bottom, and the receiving water salinity, density, and velocity. Receiving water salinity, density, and velocity may vary with water depth, distance from the discharge point and time of day and year.

Discharge conditions include the discharge geometry, and the discharge flow conditions. The discharge geometry can vary from the end of the pipe to a lengthy multi-port diffuser. The discharge can be at the water surface or submerged. The submerged outfall can be buried (except for ports) or not. Much of the historical outfall design work deals with discharges from WWTPs. These discharges can be very large—up to several hundred MGDs in flow. In ocean outfalls and in many inland outfalls, these discharges are of lower salinity than the receiving water, and the discharge has positive buoyancy. The less dense effluent rises in the denser receiving water after it is discharged.

The volume of flow of membrane concentrates is on the lower side of the range of WWTP effluent volumes, extending up to perhaps 15 MGD at present. Membrane concentrate, as opposed to WWTP effluent, tends to be of higher salinity then most receiving waters, resulting in a condition of negative buoyancy where the effluent sinks after it is discharged. This presents a concern of the potential impact of the concentrate on the benthic community at the receiving water bottom. Any possible effect on the benthic community is a function of the local ecosystem, the composition of the discharge, and the degree of dilution present at the point of contact. The chance of an adverse impact is reduced by increasing the amount of dilution at the point of bottom contact through diffuser design.

With regard to concentrate discharge regulations, it is important to note that receiving waters can differ substantially in their volume, flow, depth, temperature, composition, and degree of variability in these parameters. The effect of discharge of a concentrate or backwash to a receiving water can vary widely depending on these factors. The regulation of effluent disposal to receiving water involves several considerations, some of which are the end-of-pipe characteristics of the concentrate or backwash. Comparison is made between receiving water quality standards (dependent on the classification of the receiving water) and the water quality of the effluent to determine disposal feasibility. In addition, in states such as Florida, the effluent must also pass a test where test species (chosen based on the receiving water characteristics) are exposed to various dilutions of the effluent. Because the nature of the concentrate or backwash is different than that of the receiving water, there is a region near the discharge area where mixing and subsequent dilution of the concentrate or backwash occurs.

Where conditions cannot be met at the end of the discharge pipe, a mixing zone may be granted by the regulatory agency. The mixing zone is an administrative construct that defines a limited area or volume of the receiving water where this initial dilution of the discharge can occur. The definition of an allowable mixing zone is based on receiving water modeling. The regulations require that certain conditions be met at the edge of the mixing zone in terms of concentration and toxicity.

Once the mixing zone conditions are met, then the outfall structure can be properly designed and installed. The purpose of the outfall structure is to assure that mixing conditions can be met and that discharge of the effluent, in general, will not produce any damaging effect on the receiving water, its lifeforms, wildlife, and the surrounding area.

In highly turbulent and moving receiving water with large volume relative to the effluent discharge, simple discharge from the end of a pipe may be sufficient to assure rapid dilution and mixing of the effluent. For most situations, however, the mixing can be improved substantially though the use of a carefully designed outfall structure. Such design may be necessary to meet regulatory constraints. The most typical outfall structure for this purpose consists of a pipe of limited length mounted perpendicular to the end of the delivery pipe. This pipe, called a diffuser, has one or more discharge ports along its length.

Disposal to the Sewer   Where possible, this means of disposal is simple and usually cost effective. Disposal to sewer does not require a permit but does require permission from the wastewater treatment plant. The impact of both the flow volume and composition of the concentrate will be considered by the WWTP, as it will affect their capacity buffer and their NPDES permit. The high volume of some concentrates prohibits their discharge to the local WWTP. In other cases, concerns are focused on the increased TDS level of the WWTP effluent that results from the concentrate discharge. The possibility of disposal to sewer is highly site dependent. In addition to the factors mentioned, the possibility is influenced by the distance between the two facilities, by whether the two facilities are owned by the same entry, and by future capacity increases anticipated. Where disposal to the sewer is allowed, the water treatment plant (WTP) may be required to pay fees based on volume and/ or composition.

Deep Well Disposal   As mentioned earlier, injection wells are a disposal option in which liquid wastes are injected into porous subsurface rock formations. Depths of the wells typically range from 1,000 to 8000 ft. The rock formation receiving the waste must possess the natural ability to contain and isolate it. Paramount in the design and operation of an injection well is the ability to prevent movement of wastes into or between underground sources of drinking water. Historically, this disposal option has been referred to as deep well injection or disposal to waste disposal wells. Because of the very slow fluid movement in the injection zone, injection wells may be considered a storage method rather than a disposal method; the wastes remain there indefinitely if the injection program has been properly planned and carried out.

Because of their ability to isolate hazardous wastes from the environment, injection wells have evolved as the predominant form of hazardous waste disposal in the United States. According to a 1984 study by EPA, almost 60% of all hazardous waste disposed of in 1981, or approximately 10 billion gallons, was injected into deep wells. By contrast, only 35% of this waste was disposed of in surface impoundments and less than 5% in landfills. The EPA study also found that a still smaller volume of hazardous waste, under 500 million gallons, was incinerated in 1981 (Gordon, 1984). Although RO concentrate is not classified as hazardous, injection wells are widely used for concentrate disposal in the state of Florida.

A study prepared for the Underground Injection Practices Council showed that relatively few injection well malfunctions have resulted in contamination of water supplies (Strycker and Collins, 1987). However, other studies document instances of injection well failure resulting in contamination of drinking water supplies and groundwater resources (Gordon, 1984).

Injection of hazardous waste can be considered safe if the waste never migrates out of the injection zone. However, there are at least five ways that water may migrate and contaminate potable groundwater (Strycker and Collins, 1987). Wastes may:

- Escape through the well bore into an underground source of drinking water because of insufficient casing or failure of the injection well casing due to corrosion or excessive injection pressure.
- Escape vertically outside of the well casing from the injection zone into an underground source of drinking water (USDW) aquifer.

- Escape vertically from the injection zone through confining beds that are inadequate because of high primary permeability, solution channels, joints, faults, or induced fractures.
- Escape vertically from the injection zone through nearby wells that are improperly cemented or plugged or that have inadequate or leaky casing.
- Contaminate groundwater directly by lateral travel of the injected wastewater from a region of saline water to a region of freshwater in the same aquifer.

Evaporation Pond Disposal   Solar evaporation, a well-established method for removing water from a concentrate solution, has been used for centuries to recover salt (sodium chloride) from seawater. There are also installations that are used for the recovery of sodium chloride and other chemicals from strong brines, such as the Great Salt Lake and the Dead Sea, and for the disposal of brines resulting from oil well operation (Office of Saline Water, 1970).

Evaporation ponds for membrane concentrate disposal are most appropriate for smaller volume flows and for regions having a relatively warm, dry climate with high evaporation rates, level terrain, and low land costs. These criteria apply predominantly in the western half of the United States--in particular, the southwestern portion. Advantages associated with evaporation ponds are described in the following list:

- They are relatively easy and straightforward to construct.
- Properly constructed evaporation ponds are low maintenance and require little operator attention compared to mechanical equipment.
- Except for pumps to convey the wastewater to the pond, no mechanical equipment is required.
- For smaller volume flows, evaporation ponds are frequently the least costly means of disposal, especially in areas with high evaporation rates and low land costs.

Despite the inherent advances of evaporation ponds, they are not without disadvantages that can limit their application, as described in the following list:

- They can require large tracts of land if they are located where the evaporation rate is low or the disposal rate is high.
- Most states require impervious liners of clay or synthetic membranes such as polyvinylchloride (PVC) or Hypalon. This requirement substantially increases the costs of evaporation ponds.
- Seepage from poorly constructed evaporation ponds can contaminate underlying potable water aquifers.
- There is little economy of scale (i.e., no cost reduction resulting from increased production) for this land-intensive disposal option. Consequently, disposal costs can be large for all but small-size membrane plants.

In addition to the potential for contamination of groundwater, evaporation ponds have been criticized because they do not recover the water evaporated from the pond. However, the water evaporated is not "lost"; it remains in the atmosphere for about 10 days and then returns to the surface of the earth as rain or snow. This hydrologic cycle of evaporation and condensation is essential to life on land and is larger responsible for weather and climate.

With regard to evaporation pond design considerations, sizing of the ponds, determination of the evaporation rate, and pond depth are important. Again, evaporation ponds function by transferring liquid water in the pond to water vapor in the atmosphere about the pond. The rate at which an evaporation pond can transfer this water governs the size of the pond. Selection of pond size requires determination of both the surface area and the depth needed. The surface area required is dependent primarily on the evaporation rate. The pond must have adequate depth for surge capacity and water storage, storage capacity for precipitated salts, and freeboard for precipitation (rainfall) and wave action.

Proper sizing of an evaporation pond depends on accurate calculation of the annual evaporation rate. Evaporation from a freshwater body, such as a lake, is dependent on local climatological conditions, which are very site specific. To develop accurate evaporation data throughout the United States, meteorological stations have been established at which special pans simulate evaporation from large bodies of water such as lakes, reservoirs, and evaporation ponds. The pans are fabricated to standard dimensions and are situated to be as representative of a natural body of water as possible. A standard evaporation pan is referred to as a Class A pan. The standardized dimensions of the pass and the consistent methods for collecting the evaporation data allow comparatively and reasonably accurate data to be developed for the United States. The data collection must cover several years to be reasonably accurate and representative of site-specific variations in climatic conditions. Published evaporation rate databases typically cover a 10-year or more periods and are expressed in inches per year.

The pan evaporation data from each site can be compiled into a map of pan evaporation rates. Because of the small heat capacity of evaporation ponds, they tend to heat and cool more rapidly than adjacent lakes and to evaporate at a higher rate than an adjacent natural pond of water. In general, experience has shown the evaporation rate from large bodies of water to be approximately 70% of that measured in a Class A pan (Reclamation, 1969). This percentage is referred to as the Class A pan coefficient and must be applied to measured pan evaporation to arrive at actual lake evaporation. Over the years, site-specific Class A pan coefficients have been developed for the entire United States. Multiplying the pan evaporation rate by the pan coefficient results in a mean annual lake evaporation rate for a specific area.

Maps depicting annual average precipitation across the United States also are available. Subtracting the mean annual evaporation from the mean annual precipitation gives the net lake surface evaporation in inches per year. This is the amount of water that will evaporate from a freshwater pond (or the amount the surface level will drop) over a year if no water other than natural precipitation enters the pond. All these maps assume an impervious pond that allows no seepage. Note that for some parts of the country, the results of this calculation give a negative number; and in other parts of the country, it is a positive number. A negative number indicates a net loss of water from a pond over a year, or a drop in the pond surface level. A possible number indicates more precipitation than evaporation at a particular site. A freshwater pond at one of these sites would actually gain water over a year, even if no water other than natural precipitation were added. Thus, such a site would not be a candidate for an evaporation pond.

DID YOU KNOW?

RO concentrate streams are not easily disposed of in inland areas, as surface water and sanitary sewer discharges would not be allowed, and deep well injection may not be feasible depending on geologic features.

It is important to realize that data of this type are representative only of the particular sites of the individual meteorological stations, which may be separated by many miles. Climatic data specific to the exact site should be obtained if at all possible before actual construction of an evaporation pond.

The evaporation data described above are for freshwater pond evaporation. However, brine density has a marked effect on the rate of solar evaporation. Most procedures for calculating evaporation rate indicate evaporation is directly proportional to vapor pressure. Salinity reduces evaporation primarily because the vapor pressure of the saline water is lower than that of freshwater and because dissolved salts lower the free energy of the water molecules. Cohesive forces acting between the dissolved ions and the water molecules may also be responsible for inhibiting evaporation, making it more difficult for the water to escape as vapor (Miller, 1989).

The lower vapor pressure and lower evaporation rate of saline water result in a lower energy loss and, thus, a higher equilibrium temperature than that of freshwater under the same exposure conditions. The increase in temperature of the saline water would tend to increase evaporation, but the water is less efficient in converting radiant energy into latent heat due to the exchange of sensible heat and longwave radiation with the atmosphere. The net result is that, with the same input of energy, the evaporation rate of saline water is lower than that of freshwater. To provide the non-engineer or non-scientist with an idea of how (for illustrative reasons only) evaporation pond evaporation rate is mathematically determined, the following is provided.

Estimating Rate of Evaporation Pond Evaporation Rate

In lake, reservoir, and pond management, knowledge of evaporative processes is important to the environmental professional in understanding how water losses through evaporation are determined. Evaporation increases the storage requirement and decreases the yield of lakes and reservoirs. Several models and empirical methods are used for calculating lake and reservoir evaporative processes. In the following, applications used for the water budget and energy budget models, along with four empirical methods: The Priestly-Taylor, Penman, DeBruin-Keijman, and Papadakis equations are made.

Water Budget Model

The water budget model for lake evaporation is used to make estimations of lake evaporation in some areas. It depends on an accurate measurement of the inflow and outflow of the lake. It is expressed as

$$\Delta S = P + R + GI - GO - E - T - O \tag{12.5}$$

where

ΔS = change in lake storage, mm
P = precipitation, mm
R = surface runoff or inflow, mm
GI = groundwater inflow, mm
GO = groundwater outflow, mm
E = evaporation, mm
T = transpiration, mm
O = surface water release, mm

If a lake has little vegetation and negligible groundwater inflow and outflow, lake evaporation can be estimated by:

$$E = P + R - O \pm \Delta S \tag{12.6}$$

Note: Much of the following information is adapted from Mosner and Aulenbach (2003) *Comparison of Methods used to Estimate Lake Evaporation for a Water Budget of Lake Seminole, Southwestern Georgia, and Northwestern Florida.* Atlanta, Georgia.

Energy Budget Model

According to Rosenberry et al. (1993), the energy budget (Lee and Swancar, 1996) is recognized as the most accurate method for determining lake evaporation. Mosner and Aulenbach (2003), point out that it is also the most costly and time-consuming method. The evaporation rate, E_{EB}, is given by (Lee and Swancar, 1996)

$$E_{EB}, \text{cm/day} = \frac{Q_s - Q_r + Q_a + Q_{ar} - Q_{bs} + Q_v - Q_x}{L(1 + BR) + T_0} \tag{12.7}$$

where

E_{EB} = evaporation, in centimeters per day (cm/day)
Q_s = incident shortwave radiation, in cal/cm^2/day
Q_r = reflected shortwave radiation, in cal/cm^2/day
Q_a = incident longwave radiation from atmosphere, in cal/cm^2/day
Q_{ar} = reflected longwave radiation, in cal/cm^2/day

Q_{bs} = longwave radiation emitted by lake, in cal/cm²/day
Q_v = net energy advected by streamflow, ground water, and precipitation, in cal/cm²/day
Q_x = change in heat stored in water body, in cal/cm²/day
L = latent heat of vaporization, in cal/g
BR = Bowen Ratio, dimensionless
T_0 = water-surface temperature (°C)

Priestly–Taylor Equation

Winter and colleagues (1995) point out that the Priestly–Taylor equation is used to calculate potential evapotranspiration (PET) which is a measure of the maximum possible water loss from an area under a specified set of weather conditions or evaporation as a function of latent heat of vaporization and heat flux in a water body, and is defined by the equation:

$$PET, cm \: / \: day = \alpha \left(s \: / \: s + \gamma \right) \left[\left(Q_n - Q_x \right) / L \right] \qquad (12.8)$$

where
PET = potential evapotranspiration, cm/day
α = 1.26, Priestly–Taylor empirically derived constant, dimensionless
$(s/s + \gamma)$ = parameters derived from slope of saturated vapor pressure–temperature curve at the mean air temperature; γ is the psychrometric constant, s is the slope of the saturated vapor pressure gradient, dimensionless
Q_n = net radiation, cal/cm²/day
Q_x = change in heat stored in water body, cal/cm²/day
L = latent heat of vaporization, cal/g

Penman Equation

Winter et al. (1995) point out that the Penman equation for estimating potential evapotranspiration, E_0, can be written as:

$$E_0 = \frac{\left(\Delta / \gamma \right) H_e + E_a}{\left(\Delta / \gamma \right) + 1} \qquad (12.9)$$

where
Δ = slope of the saturation absolute humidity curve at the air temperature
γ = the psychrometric constant
H_e = evaporation equivalent of the net radiation
E_a = aerodynamic expression for evaporation

DeBruin–Keijman Equation

The DeBruin–Keijman equation (Winter et al., 1995) determines evaporation rates as a function of the moisture content of the air above the water body, the heat stored in the still water body, and the psychrometric constant, which is a function of atmospheric pressure and latent heat of vaporization.

$$PET, cm/day = \left[SVP/0.95SVP + 0.63\gamma)\right]*\left(Q_n - Q_x\right) \qquad (12.10)$$

where
SVP = saturated vapor pressure at mean air temperature, millibars/K
All other terms have been defined previously

Papadakis Equation

The Papadakis equation (Winter et al., 1995) does not account for the heat flux that occurs in the still water body to determine evaporation. Instead, the equation depends on the difference in the saturated vapor pressure above the water body at maximum and minimum air temperatures, and evaporation is defined by the equation

$$PET, cm/day = 0.5625\left[E_0 max - \left(E_0 min - 2\right)\right] \qquad (12.11)$$

where all terms have been defined previously.

For water saturated with sodium chloride salt (26.4%), the solar evaporation rate is generally about 70%t of the rate for freshwater (Office of Saline Water, 1971). Studies have shown that the evaporation rate from the Great Salt Lake, which as a TDS level of between 240,000 and 280,000 mg/L, is about 80–82% of the rate for freshwater. Other studies indicate that evaporation rates of 2-, 5-, 10-, and 20% sodium chloride solutions are 97, 98, 93, and 78%t, respectively, of the rates of freshwater (Reclamation,1969). These ratios are determined from both experiment and theory. However, there is no simple relationship between salinity and evaporation, for there are always complex interactions among site-specific variables such as air temperature, wind velocity, relative humidity, barometric pressure, water-surface temperature, heat exchange rate with the atmosphere, including incident solar absorption and reflection, thermal currents in the pond, and depth of the pond. As a result, these ratios should be used only as guidelines and with discretion. It is important to recognize that salinity can significantly reduce evaporation rate and to allow for this effect in sizing the evaporation ponds surface area. In lieu of site-specific data, an evaporation ratio of 0.70 is a reasonable allowance for long-term evaporation reduction. This ratio is also considered to be an appropriate factor for evaporation ponds that are expected to reach salt saturation over their anticipated service life.

DID YOU KNOW?

To gain understanding of what is meant by incident solar absorption, the following definitions are provided:

Incident Ray—a ray of light that strikes (impinges upon) a surface. The angle between this ray and the perpendicular or normal to the surface is the angle of incidence.
Reflected Ray—a ray that has rebounded from a surface.
Angle of Incidence—the angle between the incident ray and a normal line.

Angle of Reflection—the angle between the reflected ray and the normal line.
Angle of Refraction—the angle between the refracted ray and the normal line.
Index of Refraction (n)—is the ratio speed of light (c) in a vacuum to its speed (v) in a given material; always greater than 1.

As mentioned earlier, pond depth is an important parameter in determining pond evaporation rate. Studies indicated that pond depths ranging from 1 to 18 inches are optimal for maximizing evaporation rate. However, similar studies indicate only a 4% reduction in the evaporation rate as the pond depth is increased from 1 to 40 inches (Reclamation, 1969). Very shallow evaporation ponds are subject to drying and cracking of the liners and are not functional in long-term service for concentrate disposal. From a practical operating standpoint, an evaporation pond must not only evaporate wastewater but also provide

- Surge capacity or contingency water storage.
- Storage capacity for precipitated salts.
- Freeboard for precipitation and wave action.

For an evaporation pond to be a viable disposal alternative for membrane concentrate it must be able to accept concentrate at all times and under all conditions so as not to restrict operation of the desalination plant. The pond must be able to accommodate variations in the weather and upsets in the desalination plant. The desalination plant cannot be shut down because the evaporation pond level is rising faster than anticipated.

DID YOU KNOW?

Current concentrate disposal of membrane concentrate using evaporation ponds accounts for 5% of total disposal practices.

To allow for unpredictable circumstances, it is important that design contingencies be applied to the calculated pond area and depth. Experience from the design of industrial evaporation ponds has shown that discharges are largest during the first year of plant operation, are reduced during the second year, and are relatively constant thereafter. A long-term, 20% contingency may be applied to the surface areas of the pond or its capacity to continuously evaporate water. The additional contingencies above the 20% (up to 50%) during the first and second years of operation are applied to the depth holding capacity of the pond.

Freeboard for precipitation should be estimated on the basis of precipitation intensity and duration for the specific site. There may also be local codes governing freeboard requirements. In lieu of site-specific data, an allowance of 6 inches for precipitation is generally adequate where evaporation ponds are most likely to be located in the United States (Office of Saline Water, 1970).

Freeboard for wave action can be estimated as follows (Office of Saline Water, 1970):

$$Hw = 0.047 \times W \times \sqrt{(F)}$$ (12.12)

where:
Hw = wave height (ft)
W = wind velocity (mph)
F = fetch, or straight-line distance the wind can blow without obstruction (mi)

The run-up of waves on the face of the dike approaches the velocity head of the waves and can be approximated as 1.5 x wave height (Hw). *Hw* is the freeboard allowance for wave action and typically ranges from 2 to 4 ft. The minimum recommended combined freeboard (for precipitation and wave action) is 2 ft. This minimum applies primarily to small ponds.

Over the life of the pond (which should be sized for the same duration as the projected life of the desalination facility), the water will likely reach saturation and precipitate salts. The type and quantity of salts is highly variable and very site specific.

Spray Irrigation Disposal

Land application methods include irrigation systems, rapid infiltration, and overland flow systems (Crites et al., 2000). These methods, and in particular irrigation, were originally used to take advantage of sewage effluent as a nutrient or fertilizer source as well as to reuse the water. Membrane concentrate has been used for land application in the spray irrigation mode. Using the concentrate in lieu of fresh irrigation water helps conserve natural resources, and in areas where water conservation is of great importance, spray irrigation is especially attractive. Because of the higher TDS concentration of RO concentrate, unless it is diluted, concentrate is less likely to be used for spray irrigation purposes.

DID YOU KNOW?

The removal of nutrients is one advantage spray irrigation has compared to conventional disposal methods like instream discharge.

Concentrate can be applied to cropland or vegetation by sprinkling or surface techniques for water conservation by exchange when lawns, parks, or golf course are irrigated and for preservation and enlargement of green belts and open spaces.

Where the nutrient concentration of the wastewater for irrigation is of little value, hydraulic loading can be maximized to the extent possible, and system costs can be minimized. Crops such as water-tolerant grasses with low potential for economic return but with high salinity tolerance are generally chosen for this type of requirement.

Fundamental considerations in land application systems include knowledge of wastewater characteristics, vegetation, and public health requirements for successful design and operation. Environmental regulations at each site must be closely examined to determine if spray irrigation is feasible. Contamination of the groundwater and runoff into surface water are key concerns. Also, the quality of the concentrate—its salinity, toxicity, and the soil permeability—must be acceptable.

The principal objective in spray irrigation systems for concentrate discharge is ultimate disposal of the applied wastewater. With this objective, the hydraulic loading is usually limited by the infiltration capacity of the soil. If the site has a relatively impermeable subsurface layer or a high groundwater table, underdrains can be installed to increase the allowable loading. Grasses are usually selected for the vegetation because of their high nutrient requirements and water tolerance.

Other conditions must be met before concentrate irrigation can be considered as a practical disposal option. First, there must be a need for irrigation water in the vicinity of the membrane plant. If the need exists, a contract between the operating plant and the irrigation user would be required. Second, a backup disposal or storage method must be available during periods of heavy rainfall. Third, monitor wells must be drilled before an operating permit is obtained (Conlon, 1989).

With regard to design factors, the following considerations are applicable to spray irrigation of concentrate for ultimate disposal:

- Salt, trace metals, and salinity
- Site selection
- Preapplication treatment
- Hydraulic loading rates
- Land requirements
- Vegetation selection
- Distribution techniques.
- Surface runoff control

Salt, Trace Metals, and Salinity

Three factors that affect an irrigation source's long-term influence on soil permeability are the sodium content relative to calcium and magnesium, the carbonate and bicarbonate content, and the total salt concentration of the irrigation water. Sodium salts remain in the soil and may adversely affect its structure. High sodium concentrations in clay-bearing soils disperse soil particles and decrease soil permeability, thus reducing the rate at which water moves into the soil and reducing aeration. If the soil permeability, or infiltration rate, is greatly reduced, then the vegetation on the irrigation site cannot survive. The hardness level (calcium and magnesium) will form insoluble precipitates with carbonates when the water is concentrated. This buildup of solids can eventually block the migration of water through the soil.

The US Department of Agriculture's Salinity Laboratory developed a sodium adsorption ratio (SAR) to determine the sodium limit. It is defined as follows:

$$SAR = Na/\left[Ca + Mg \right)/2]^{1/2} \tag{12.13}$$

where
    Na = sodium, milliequivalent per liter (meq/L)
    Ca = calcium, meq/L
    Mg = magnesium, meq/L

High SAR values (>9) may adversely affect the permeability of fine-textured soils and can sometimes be toxic to plants.

Trace elements are essential for plant growth; however, at higher levels, some become toxic to both plants and microorganisms. The retention capacity for most metals in most soils is generally high, especially for pH above 7. Under low pH conditions, some metals can leach out of soils and may adversely affect the surface waters in the area.

Salinity is the most important parameter in determining the impact of the concentrate on the soil. High concentrations of salts whose accumulation is potentially harmful will be continually added to the soil with irrigation water. The rate of salt accumulation depends upon the quantity applied and the rate at which it is removed from the soil by leaching. The salt levels in many brackish reverse osmosis concentrates can be between 5,000 and 10,000 parts per million, a range that normally rules out spray irrigation.

DID YOU KNOW?

Soluble salts in a water solution will conduct an electric current; thus, changes in electrical conductivity (EC) can be used to measure the water's salt content in electrical resistance units (decisiemens per meter, or dS m^{-1}),

In addition to the effects of total salinity on vegetation and soil, individual ions can cause reduction in plant growth. Toxicity occurs when a specific ion is taken up and accumulated by the vegetation, ultimately resulting in damage to it. The ions of most concern in wastewater effluent irrigation are sodium, chloride, and boron. Other heavy metals can be very harmful, even if present only in small quantities. These include copper, iron, barium, lead, and manganese. These all have strict environmental regulations in many states.

In addition to the influence on the soil, the effect of the salt concentrations on the groundwater must be considered. The possible impact on groundwater sources may be a difficult obstacle where soil saturation is high, and the water table is close to the surface. The chance of increasing background TDS levels of the groundwater is high with the concentrate. Due to this consideration, spray irrigation requires a runoff control system. An underdrain or piping distribution system may have to be installed under the full areas of irrigation to collect excess seepage through the soil and, thus, to protect the groundwater sources. If high salinity concentrate is being used, scaling of the underdrain may become a problem. The piping perforations used to collect the water can be easily scaled because the openings are generally small. Vulnerability to scaling must be carefully evaluated before a project is undertaken.

TABLE 12.7
Site Selection Factors and Criteria

Factor	Criterion
Soil	
Type	Loamy soils are preferred, but most soils from sands to clays are acceptable.
Drainability	Well-drained soil is preferred.
Depth	Uniformly 5 to 6 feet or more throughout sites is preferred.
Groundwater	
Depth to groundwater	A minimum of 5 feet is preferred.
Groundwater control	Control may be necessary to ensure renovation if the water table is less than 10 feet from the surface.
Groundwater movement	Velocity and direction of movement must be determined.
Slopes	Slopes of up to 20% are acceptable with or without terracing.
Underground formations	Formations should be mapped and analyzed with respect to interference with groundwater or percolating water movements.
Isolation	Moderate isolation from public is preferred, the degree of isolation depends on wastewater characteristics, methods of application, and crop.
Distance from source of wastewater	an appropriate distance is a matter of economics.

Site Selection

Site selection factors and criteria for effluent irrigation are presented in Table 12.7. A moderately permeable soil capable of infiltration up to 2 inches per day on an intermittent basis is preferable. The total amount of land required for land application is highly variable but primarily depends on application rates.

Preapplication Treatment

Factors that should be considered in assessing the need for preapplication treatment include whether the concentrate is mixed with additional wastewaters before application, the type of vegetation grown, the degree of contact with the wastewater by the public, and the method of application. In four Florida sites, concentrate is aerated before discharge, because each plant discharges to a retention pond or ponds before irrigation. Aeration by increasing DO prevents stagnation and algae growth in the ponds and also supports fish populations. The ponds are required for flow equalization and mixing. Typically, concentrate is blended with biologically treated wastewater.

Hydraulic Loading Rates

Determining the hydraulic loading rate is the most critical step in designing a spray irrigation system. The loading rate is used to calculate the required irrigation area and is a function of precipitation, evapotranspiration, and percolation. The following equation represents the general water balance for hydraulic loading based upon a monthly time period and assuming zero runoff:

$$HLR = ET + PER - PPT \qquad (12.14)$$

where:
   HLR = hydraulic loading rate
   ET = evapotranspiration
   PER = percolation
   PPT = precipitation

In most cases, surface runoff from fields irrigated with wastewater is not allowed without a permit or, at least, must be controlled; it is usually controlled just so that a permit does not have to be obtained.

Seasonal variations in each of these values would be considered by evaluating the water balance for each month as well as the annual balance. For precipitation, the wettest year in 10 is suggested as reasonable in most cases. Evapotranspiration will also vary from month to month, but the total for the year should be relatively constant. Percolation includes that portion of the water that, after infiltration into the soil, flows through the root zone and eventually becomes part of the groundwater. The percolation rate used in the calculation should be determined on the basis of a number of factors, including soil characteristics underlying geologic conditions, groundwater conditions, and the length of drying period required for satisfactory vegetation growth. The principal factor is the permeability of hydraulic conductivity of the least permeable layer in the soil profile.

Resting periods, standard in most irrigation techniques, allow the water to drain from the top few inches of soil. Aerobic conditions are thus restored, and air penetrates the soil. Resting periods may range from a portion of each day to 14 days and depend on the vegetation, the number of individual plots in the rotation cycle, and the availability of backup storage capacity.

To properly calculate an annual hydraulic loading rate, monthly evapotranspiration, precipitation, and percolation rates must be obtained. The annual hydraulic loading rate represents the sum of the monthly loading rates. Recommended loading rates range from 2 to 20 ft per year (Goigel, 1991).

Land Requirements

Once a hydraulic loading rate has been determined, the required irrigation area can be calculated using the following equation:

$$A = Q \times Kl/ALR \qquad (12.15)$$

where:
   A = irrigation area (acre)
   Q = concentrate flow (gpd)
   ALR = annual hydraulic loading rate (ft/yr)
   Kl = 0.00112 d x ft^3 x acres/(hr x gal x ft^2)

The total land area required for spray irrigation includes allowances for buffer zones and storage and, if necessary, land for emergencies or future expansion.

For loadings of constituents such as nitrogen, which may be of interest to golf course managers who need fertilizer for the grasses, the field area requirement is calculated as follows:

$$\text{Field Area}(\text{acres}) = 3,040 \times C \times Q / Lc \qquad (12.16)$$

where:
C = concentration of constituent (mg/L)
Q = flow rate (mgd)
Lc = loading rate of constituent (pounds per acre-year [lb/acre-yr])

Vegetation Selection
The important aspects of vegetation for irrigation systems are water needs and tolerances, sensitivity to wastewater constituents, public health regulations, and vegetation management considerations. The vegetation selection depends highly on the location of the irrigation site and natural conditions such as temperature, precipitation, and topsoil condition. Automated watering alone cannot always ensure vegetation propagation. Vegetation selection is the responsibility of the property owners. Woodland irrigation for growing trees is being conducted in some areas. The principal limitations on this use of wastewater include low water tolerances of certain trees and the necessity to use fixed sprinklers, which are expensive.

Membrane concentrate disposal will generally be to landscape vegetation. Such application, for example to highway median and border strips, airport strips, golf courses, parks and recreational areas, and wildlife areas, has several advantages. Problems associated with crops for consumption are avoided, and the irrigated land is already owned, so land acquisition costs are saved.

Distribution Techniques
Many different distribution techniques are available for engineered wastewater effluent applications. For irrigation, two main groups, sprinkling and surface application, are used. Sprinkling systems used for spray irrigation are of two types—fixed and moving. Fixed systems, often called solid set systems, may be either on the ground surface or buried. Both types usually consist of impact sprinklers mounted on risers that are spaced along lateral pipelines, which are, in turn, connected to main pipelines. These systems are adaptable to a wide variety of terrains and may be used for irrigation of either cultivated land or woodlands. Portable aluminum pipe is normally used for aboveground systems. This pipe has the advantage of relatively low capital cost but is easily damaged, has a short, expected life because of corrosion, and must be removed during cultivation and harvesting operations. Pipe used for buried systems may be buried as deep and 1.5 ft below the ground surface. Buried systems usually have the greatest capital cost; however, they are probably the most dependable and are well suited to automated control.

There are a number of different moving sprinkle systems, including center-pivot, side-roll, wheel-move, rotating-boom, and winch-propelled systems.

Surface Runoff Control

Surface runoff control depends mainly on the proximity of surface water. If runoff drains to surface water, an NPDES permit may be required. This situation should be avoided if possible due to the complication of quantifying overland runoff. Berms can be built around the irrigation field to prevent runoff. Another alternative, although expensive, is a surrounding collection system. It is best to use precautions and backup systems to ensure that overwatering and subsequent runoff do not occur in the first place.

Zero Liquid Discharge Disposal

In this approach, evaporation is used to further concentrate the membrane concentrate. In the extreme limit of processing concentrate to dry salts, the method becomes a zero-discharge option. Evaporation requires major capital investment, and the high energy consumption together with the final salt or brine disposal can result in significant disposal costs. Because of this, disposal of municipal membrane concentrate by mechanical evaporation would typically be considered as a last resort; that is, when no other disposal option is feasible. Cost aside, however, there are some advantages:

- It may avoid a lengthy and tedious permitting process.
- It may gain quick community acceptance.
- It can be located virtually anywhere.
- It represents a positive extreme in recycling by efficiently using the water source.

When this thermal process is used following an RO system, for example, it produces additional product water by recovering high-purity distillate from the concentrate wastewater stream. The distillate can be used to help meet the system product water volume requirement. This reduces the size of the membrane system and, thus, the size of the membrane concentrate to be treated by the thermal process. In addition, because the product purity of the thermal process is so high (TDS in the range of 10 mg/L), some of the product water volume reduction of the system may be met by blending the thermal product with untreated source water. The usual concerns and consideration using untreated water for blending need to be addressed. The end result may be a system where the system product requirement is met by three streams: (1) membrane product, (2) thermal process product, and (3) bypass water.

Single- and Multiple-Effect Evaporators

Using steam as the energy source, it takes about 1,000 British thermal units (Btu) to evaporate a pound of water. In a single-effect evaporator, heat released by the condensing steam is transferred across a heat exchange surface to an aqueous solution boiling at a temperature lower than that of the condensing steam. The solution absorbs heat; and part of the solution water vaporizes, causing the remaining solution to become richer in solution. The water vapor flows to a barometric or surface condenser, where it condenses as its latent heat is released to cooling water at a lower temperature. The finite temperature differences between the steam, the boiling liquid, and the condenser are the driving forces required for the heat transfer surface area to be less than infinite. Practically all the heat removed from the condensing steam (which had been generated initially by burning fuel) is rejected to cooling water and is often dissipated to the environment without being of further use.

The water vapor that flows to the condenser in a single-effect evaporator is at a lower temperature and pressure than the heating steam but has almost as much enthalpy. Instead of releasing its latent heat to cooling water, the water vapor may be used as heating steam in another evaporator effect, operating at a lower temperature and pressure than the first effect. Additional effects may be added in a similar manner, each generating additional vapor, which may be used to heat a lower-temperature effect. The vapor generated in the lowest-temperature effect finally is condensed by releasing its latent heat to cooling water in a condenser. The economy of a single- or multiple-effect evaporator may be expressed as the ratio of kilograms of total evaporation to kilograms of heating steam. As effects are added, the economy increases, representing more efficient energy utilization. Eventually, added effects result in marginal added benefits, and the number of effects is thus limited by both practical and economic considerations. Multiple-effect evaporators increase the efficiency (economy) but add capital cost in additional evaporator bodies.

More specifically, the number of effects, and thus the economy achieved, is limited by the total temperature difference between the saturation temperature of the heating steam (or heat source) and the temperature of the cooling water (or other heat sink). The available temperature difference may also be constrained by the temperature sensitivity of the solution to be evaporated. The total temperature difference, less any losses, becomes allocated between effects in proportion to their resistance to heat transfer, the effects being thermal resistances in series.

The heat transfer surface area for each effect is inversely proportional to the net temperature difference available for that effect. Increasing the number of effects reduces the temperature difference and evaporation duty per effect, which increases the total area of the evaporator in rough proportion to the number of effects. The temperature difference available to each effect is reduced by boiling point elevation and by the decrease in vapor saturation temperature due to pressure drop. The boiling point elevation of a solution is the increase in boiling point of the solution compared to the boiling point of pure water at the same pressure; it depends on the nature of the solute and increases with increasing solute concentration. In a multiple-effect evaporator, the boiling point elevation and vapor pressure drop losses for all the effects must be summed and subtracted from the overall temperature difference between heat source and sink to determine the net driving force available for heat transfer.

Vapor-Compression Evaporator Systems (Brine Concentrators)

A vapor-compression evaporator system, or brine concentrator, is similar to a conventional single-effect evaporator, except that the vapor released from the boiling solution is compressed in a compressor. Compression raises the pressure and saturation temperature of the vapor so that it may be returned to the evaporator steam chest to be used as heating steam. The latent heat of the vapor is used to evaporate more water instead of being rejected to cooling water. The compressor adds energy to the vapor to raise its saturation temperature above the boiling temperature of the solution by whatever net temperature difference is desired. The compressor is not completely efficient, having small losses due to mechanical friction and larger losses due to nonisentropic compression. However, the additional energy required because of nonisentropic compression is not lost from the evaporator system; it serves to superheat the compressed vapor. The compression energy added to the vapor is of the same magnitude as energy required to raise feed to the boiling point and make up for radiation and venting losses. By exchanging heat between the condensed vapors (distillate) and the product with the feed, it is usually possible to operate with little or no makeup heat in addition to the energy necessary to drive the compressor. The compressor power is proportional to the increase in saturation temperature produced by the compressor. The evaporator design must trade off compressor power consumption versus heat transfer surface area. Using the vapor-compression approach to evaporate water requires only about 100 Btu to evaporate a pound of water. Thus, one evaporator body driven by mechanical vapor compression is equivalent to 10 effects or a 10-body system driven by steam.

DID YOU KNOW?

In the British system of units, the unit of heat is the British thermal unit, or Btu. One *Btu* is the amount of heat required to raise one pound of water one degree Fahrenheit at normal atmospheric pressure (1 atm).

Although most brine concentrators have been used to process cooling water, concentrators have also been used to concentrate reject from RO plants. Approximately 90% of these concentrators operate with a seeded slurry process that allows the reject to be concentrated as much as 40 to 1 without scaling problems developing in the evaporator. Brine concentrators also produce distilled product water that can be used for high-purity purposes or for blending with other water supplies. Because of the ability to achieve such high levels of concentration, brine concentrators can reduce or eliminate the need for alternative disposal methods such as deep well injection or solar evaporation ponds. When operated in conjunction with crystallizers or spray dryers, brine concentrators can achieve zero liquid discharge of RO concentrate under all climatic conditions.

Individual brine concentrator units' range in capacity from approximately 10–700 gpm of feedwater flow. Units below 150 gpm of capacity are usually skid mounted, and larger units are field fabricated. A majority of operating brine concentrators are single-effect, vertical tube, falling film evaporators that use a calcium sulfate-seeded slurry process. Energy input to the brine concentrator can be provided by an electric-driven vapor compressor or by process steam from a host industrial facility. Steam-driven systems can be configured with multiple effects to minimize energy consumption.

Product water quality is normally less than 10 mg/L TDS. Brine reject from the concentrator typically ranges between 2 and 10% of the feedwater flow, with TDS concentrations as high as 250,000 mg/L.

DID YOU KNOW?

Solids in water occur either in solution or in suspension and are distinguished by passing the water sample through a glass-fiber filter. By definition, the *suspended solids* are retained on top of the filter, and the *dissolved solids* pass through the filter with the water. When the filtered portion of the water sample is placed in a small dish and then evaporated, the solids in the water remain as residue in the evaporating dish. This material is called *total dissolved solids*, or TDS. Dissolved solids may be organic or inorganic. Water may come into contact with these substances within the soil, on surfaces, and in the atmosphere. The organic dissolved constituents of water are from the decay products of vegetation, from organic chemicals, and from organic gases. Removing these dissolved minerals, gases, and organic constituents is desirable, because they may cause physiological effects and produce aesthetically displeasing color, taste, and odors.

Because of the corrosive nature of many wastewater brines, brine concentrators are usually constructed of high-quality materials, including titanium evaporator tubes and stainless-steel vessels suitable for 30-year evaporator life. For conditions of high chloride concentrations or other more corrosive environments, brine concentrators can be constructed of materials such as AL6XN, Inconal 825, or other exotic metals to meet performance and reliability requirements.

DID YOU KNOW?

The method of evaporation selected is based on the characteristics of the RO membrane concentrate and the type of energy source to be used.

Crystallizers

Crystallizer technology has been used for many years to concentrate feed streams in industrial processes. More recently, as the need to concentrate wastewaters has increased, this technology has been applied to reject from desalination processes, such as brine concentrate evaporators, to reduce wastewater to a transportable solid. Crystallizer technology is especially applicable in areas where solar evaporation pond construction cost is high, solar evaporation rates are negative, or deep well disposal is costly, geologically not feasible, or not permitted.

Crystallizers used for wastewater disposal range in capacity from about 2 to 50 gpm. These units have vertical cylindrical vessels with heat input from vapor compressors or an available steam supply. For small systems ranging from 2 to

6 gpm, steam-driven crystallizers are more economical. Steam can be supplied by a package boiler or a process source if one is available. For larger systems, electrically driven vapor compressors are normally used to supply heat for evaporation.

Typically, the crystallizer requires a purge stream of about 2% of the feed to the crystallizer. This is necessary to prevent extremely soluble species (such as calcium chloride) from building up in the vapor body and to prevent production of dry cake solids. The suggested disposal of this stream is to a small evaporation pond. The crystallizer produces considerable solids that can be disposed of to commercial landfill.

The first crystallizers, applied to power plant wastewater disposal, experienced problem related to materials selection and process stability; but subsequent design changes and operating experience have produced reliable technology.

For RO concentrate disposal, crystallizers would normally be operated with a brine concentrator evaporator to reduce brine concentrator blowdown to a transportable solid. Crystallizers can be used to concentrate RO reject directly, but their capital cost and energy usage is much higher than for a brine concentrator of equivalent capacity.

Spray Dryers

Spray dryers provide an alternative to crystallizers for concentration of wastewater brines to dryness. Spray dryers are generally more cost effective for smaller feed flows of less than 10 gpm.

PATHOGEN REMOVAL

Various states have developed different approaches to regulating pathogen removal by indirect potable reuse plants. For example, Virginia permitted the UOSA indirect potable reuse plant and the Broad Run WRF based on achieving a nondetected concentration of *Escherichia coli* (less than 2 cfu/100 mL). Other states have taken a different approach. For example, California requires 12-log reduction of viruses and 10-log reduction of *Cryptosporidium* and *Giardia* from the raw wastewater to the advanced water treatment plant finished water.

Pathogen removal by each of the three proposed treatment trains is significant and would result in nondetectable concentrations for all indicator organisms typically used in wastewater treatment, (e.g., *E. coli*, total coliform, fecal coliform) assuming proper operation. Therefore, compliance with a UOSA-type permit or Florida's pathogen-related indirect potable reuse regulations would be met by all three proposed treatment trains. Compliance with the California regulations may be more challenging, especially for GAC-based treatment, because of the high log reduction requirements. Discussion is necessary with the Virginia regulators in order to determine how the proposed HRSD groundwater recharge project would be regulated with respect to pathogen removal.

DISINFECTION BYPRODUCTS

Excessive formation of trihalomethanes (THMs) at WWTPs is fairly common, especially for plants that provide good nitrification. Low effluent NH_3 concentrations at these plants lead to the formation of free chlorine (rather than chloramines) during chlorine disinfection, which, when reacted with bulk organics, has a propensity to form high levels of THMs. NDMA, which is another disinfection by-product, can also form in significant concentrations during the disinfection process at WWTPs, depending on the precursors in the water and the type of chlorination practiced. Little NDMA forms in the presence of free chlorine, but significant concentrations typically form in the presence of chloramines, with dichloramine resulting in more rapid formation kinetics than monochloramine. Both THMs and NDMA can be removed by AWT processes through specialized design, but a more cost-efficient approach is to prevent their formation by withdrawing the water from the WWTP prior to disinfection (upstream of the chlorine contact basin). Specific withdrawal points at each WWTP, and the potential treatment required for TTHMs and NDMA removal should be considered in the next stage of this project.

ANTICIPATED IMPROVEMENTS TO HRSD'S EXISTING WWTPS

Some operational and capital improvements to the existing WWTPs may be required, depending on the AWT train selected for implementation and the final effluent water quality produced at each WWTP. Table 12.8 shows the improvement that will likely be required. Analysis regarding WWTP improvements should be ongoing.

From the data and evaluations presented to this point, it is obvious that the bottom line is that any of the three advanced water treatment trains—RO-based, NF-based, and GAC-based—are likely viable for groundwater recharge of effluent generated from HRSD's WWTPs. Finished water quality produced by each train will be excellent with respect to pathogen and organics removal but use of the RO-based or NF-based treatment train is necessary if TDS reduction is required. Partial RO or NF treatment could be used depending on the degree of TDS reduction required. BNR improvements will be required at some of the WWTPs to reduce the TN concentration and the propensity for organic fouling membranes in the RO- and NF-based trains.

Selection of the advanced water treatment train to be implemented at each WWTP should be based on numerous factors such as finished water quality, wastewater discharge requirements, operability, sustainability, site-specific factors (e.g., space of existing infrastructure, hydraulic), and capital and operating costs (discussed later). Ultimate selection of the advanced water treatment train will also be dictated by regulatory requirements related to treatment, finished water quality, and wastewater discharge requirements that have not yet been established; therefore, engaging the appropriate regulatory agency(ies) is important during the next phase of this project. Treatment selection may also be influenced by public perception.

Because HRSD's SWIFT project is presently operational at its Nansemond Treatment but is still a work in progress, (with time for adjustment here, there, and almost anywhere) other action items that will influence advanced water treatment train selection should be performed including the following:

- Regularly sample at each WWTP for COD, sCOD, TOC, DOC, all contaminants regulated by primary MFLs, selected CECs and parameters specific to the design of RO and NF treatment (i.e., barium, strontium, fluoride, silica, alkalinity, pH).

TABLE 12.8

Required WWTP Improvements to Address AWT Operational Impacts and/or finished Water Quality Deficiency

WWTP	MF-RO-UVAOP	MF-NF-UVAOP	FS-O3-BAC-GAC-UV
Army Base	none	none	AWT Finished Water Deficiency: TDS > 750 mg/L
			Required WWTP: if possible, reduce Influent TDS to WWTP
Boar Harbor	AWT Operational Impact: Greater Membrane fouling requiring more design due to lack of BNR	AWT Operational Impact: Greater membrane fouling requiring more conservative design due to lack of BNR	AWT Finished Water Deficiency: TN > 10 mg/L; total hardness > 150 mg/L
	AWT Finished Water Deficiency: None	AWT Finished Water Deficiency: TN>10 mg/L	Required WWTP Improvements: Add nit/denit at WWTO, if determined Necessary, add softening to AWT
	Required WWTP Improvements: Add nit/denit at WWTP	Required WWTP Improvements: Add nit/denit at WWTP	
James River	None	None	None
Nansemond	None	None	None
VIP and Williamsburg	None	AWT Finished Water Deficiency: TN approaching 10 mg/L	AWT Finished Water Deficiency: TN approaching 10 mg/L; TDS > 750 mg/L; total hardness > 150 mg/L (VIP only)
		Required WWTP Improvements: improve nit/denit at WWTP	Required WWTP Improvements: Improve nit/denit at WWTP and reduce influent TDS to WWTP (VIP only), if possible, if determined necessary, add softening to AWTP (VIP only)
York River	None	None	AWT Finished Water Deficiency: Hardness > 150 mg/L
			Required WWTP or AWT Improvements: If determined necessary, add softening to AWT

HRSD/CH2M (2016)

Note:

Nit/denit = nitrification/denitrification

- Regularly measure water quality (e.g., pH, alkalinity, TDS, hardness) in numerous Potomac Aquifer product wells.
- Evaluate site-specific conditions at each WWTP that may influence AWT train selection, including site space, hydraulics, geotechnical conditions, electrical service, and use of existing infrastructure for AWT treatment.
- Conduct an industrial and commercial water quality discharge study to characterize risk and to identify chemicals of concern that may be discharged to the collection system.
- Determine potential causes for high TDS concentrations at WWTPs where effluent TDS is greater than 500 mg/L.

GEOCHEMICAL CHALLENGES FACING SWIFT PROJECT

Evaluations and analytical groundwater flow modeling revealed that HRSD will recharge between 77 and 131 MGD to the Potomac Aquifer System using over 60 injection wells with maximum capacities approaching 3.0 MGD per well, at seven WWTPs. Groundwater flow modeling revealed that the injection wells will screen combinations of two, or all three PAS aquifers zones, according to the hydrologic conditions found at individual WWTPs. Both physical and geochemical challenges can emerge while recharging clean water into aquifers composed of reactive metal-bearing minerals, and potentially unstable clay minerals, while also containing brackish native groundwater, typical conditions found in the PAS.

Physical and chemical reactions are important relative both to the well facility operation and aquifer water quality. The potential damaging effects can come from:

- Water to water mixing of injectate water with native groundwater—impacts are in the form of physical plugging pore spaces with solids or precipitated metals, reduced permeability local to the wellbore, and eventually lower injectivity.
- Interaction of the injectate/native groundwater mix with the aquifer matrix—impacts are in the form of damage to firewater sensitive clays, precipitation or dissolution of metal-bearing minerals, and potential release of metals troublesome to injection activities (iron and manganese), or water quality issues (arsenic).

REDUCTION IN INJECTIVITY

The following factors affect the injectivity of an injection well:

- Physical plugging.
- Mineral precipitation.

Physical Plugging

A well's injection rate per unit of head buildup (draw-up) in an injection well is known as its injectivity, expressed in gallons per minute per foot (gpm/ft) of draw-up (Warner and Lehr, 1981). When injection begins, the water level in the well rises as a function of the transmitting properties of the receiving aquifer, and the well's efficiency. While the transmitting character of the aquifer should remain stable over the service life of an injection well, the available head for injection will decline as the injectate recharges the aquifer. This causes the static water level to rise toward the ground surface. The well's efficiency will decrease with time, depending on the quality of the injectate, particularly to its TSS content. TSS content, which, in water, is commonly expressed as a concentration in terms of milligrams per liter and typically is described as the amount of filterable solids in a wastewater sample. More specifically, the term *solids* means any material suspended or dissolved in water and wastewater. Although normal domestic wastewater contains a very small amount of solids (usually less than 0.1%), most treatment processes are designed specifically to remove or convert solids to a form that can be removed or discharged without causing environmental harm. However, 100% removal is unlikely. Even the most purified injectate can contain small amounts of TSS. If left to accumulate in the borehole environment (wellbore), solids can clog the screen, filter pack, and aquifer proximal to the well, which reduces the well's injectivity. Injectivity reduction increase draw-up and eventually lowers the well's injection capacity (Pyne, 2005). TSS can originate from scale or dirt in piping, treatment residuals, and reactions in the injectate that result in solids precipitation. One of the more common reactions occurs when oxygen dissolved in the injectate reacts with dissolved iron or manganese, precipitating ferric or manganese oxides, turning a dissolved component of the injectate into a source of solids.

Mineral Precipitation

In addition to physical plugging, chemical reactions between the injectate and the native groundwater, or injectate and aquifer mineralogy can precipitate metal-bearing oxides and hydroxides. These reactions often come from injectate that contains dissolved oxygen (DO). Considering the relatively small surface areas around the wellbore, precipitating metal-bearing minerals can clog pore spaces, and reduce permeability and well injectivity. An important part of the research and planning involved with HRSD's SWIFT project is to determine precisely the type and composition of injectate from the RO, nanofiltration (NF), or the BAC process to estimate its potential for plugging the wellbore.

GEOCHEMICAL CONCERNS

Beyond the problems associated with physically plugging pore spaces around the borehole, several geochemical reactions can negatively affect injection well operations. These reactions include:

- Clay minerals damage.
- Metal's precipitation.
- Mineral dissolution.

Damaging Clay Minerals

The term *clay* is applied to both materials having a particle size of less than 2 micrometers (μm) (25,400 micrometers = 1 inch) and to a complex group of poorly defined hydrous silicate minerals that contain primarily aluminum, along with other cations (potassium and magnesium) according to the exact mineral species. Displaying a platy or tabular structure, clay minerals exhibit an extremely small grain size and typically adsorb water to their particle surfaces. In aquifer sand, clays occur in trace (less than 10%) amounts as components of the aquifer's interstitial spaces, coating framework particles like quartz grains, lining or filling pore spaces, or as a weathering product of feldspars.

Damaging clays occur with the disruption of their mineral structure. The damage can arise when injecting water of significantly different ionic strength than the native groundwater, a concern when injecting dilute fresh water into an aquifer containing brackish or saline native groundwater (Drever, 1988). The dilute water contains significantly less cations, and a weaker charge than brackish native groundwater. When displacing the brackish water in the diffuse-double layer between clay particles, the weaker charge can induce repulsive forces dispersing the particles, fragmenting the clay structure while mobilizing the fragments into flowing pore water. The particles can eventually accumulate in smaller pores physically plugging the pore space and reducing the permeability of the aquifer.

Damage can also arise when injectate displays differing cation chemistry than the native groundwater and the clay minerals (Langmuir, 1997). Exchanging cations can disrupt clay mineral structure particularly when their atomic radius exceeds the radius of the exchanged cation. The larger cation fragments the tabular structure, shearing off the edges of the mineral. Plate-like fragments break off the main mineral particle and migrate with flowing groundwater. Like the damage incurred by water of differing ionic strength, migrating clay fragments will brush pile in pore spaces, physically plugging passageways, and reducing aquifer permeability. Unlike the accumulation of TSS in the wellbore, formation damage by migrating clays develops in the aquifer away from the wellbore, making its removal difficult by backflushing or even invasive rehabilitation techniques.

Mineral Precipitation

Metal-bearing minerals can precipitate in the aquifer away from the well. These reactions typically occur when the injectate contains dissolved oxygen (DO) at concentrations exceeding anoxic (DO less than 1.0 milligrams per liter [mg/l]) levels but can also occur if the pH of the injectate exceed 9.0. As surface areas in the aquifer increase geometrically away from the well, mineral precipitation does not create as great a concern as the same reactions at the borehole wall.

Mineral Dissolution

Injectate reactions with minerals in the aquifer matrices can dissolve minerals leaching their elemental components (Stuyfzand, 1993). Injectate-containing DO above anoxic concentrations will react with common, reduced metal-bearing minerals like pyrite (FeS_2) and siderite ($FeCO_3$), to release iron and other metals like manganese that occupy sites in the mineral structure. Iron and manganese can precipitate as oxide and hydroxide minerals if they contact injectate-containing DO. Oxidation of arseniferous pyrite can release arsenic, creating a water quality concern in the migrating injectate (HRSD, 2016b).

WATER QUALITY AND AQUIFER MINERALOGY

Note: Because it would require an unwieldy number of permutations (63) to assess the injection of three injectate chemicals into three discreet aquifers at seven WWTPs, the chemical composition of three injectate types and native groundwater in the three PAS aquifer zones beneath the York River Treatment Plant was chosen for discussion here. The targeting of the York River WWTP for discussion that follows is not an issue and does not skew data because York River Treatment Plant exhibits effluent and local aquifer characteristics typical of conditions across the Hampton Roads Sanitation District.

Mass-balance relationships between raw water entering the plant and modeling of the advanced water treatment process were used to determine injectate chemistry. As no wells installed in the PAS aquifer zones currently exist at the York River WWTP, water quality data from the area around the site was obtained from the National Water information System (NWIS) data base (maintained by the USGS). The NWIS data base provides samples collected by USGS personnel from local municipal, irrigation, and industrial supply wells, along with designated monitoring wells.

Two methods were used to identify potential minerals in the PAS aquifers that could react with the injectate. First, thermodynamic equilibrium models were applied to identifying the potential mineral suite in each aquifer. The models were run using water chemistry analyses obtained from the NWIS database for the PAS zones around the York River WWTP area. The models project potential minerals that occur in equilibrium with water chemistry.

Secondly, mineralogical analysis of cores collected at the city of Chesapeake's ASR facility were examined to gain information on the mineralogy of the PAS. Cores were collected from the PAS zones at the city of Chesapeake. The composition of the PAS should remain fairly consistent across the HRSD service area. However, the grain size and sorting (texture) decline proceeding down the stratigraphic dip in the Virginia Coastal Plain (Treifke, 1973). Consistent with the changes in texture, the percentage of fines (texture) increase downdip. As the city of Chesapeake lies over 20 miles downdip from HRSD's York River WWTP, data from cores should portray more conservative aquifer properties than actually occur below York River.

INJECTATE WATER CHEMISTRY

Injectate chemistry was estimated by modeling water quality entering York River Treatment Plant and through the AWT processes. Effluent chemistry was estimated for reverse osmosis, nanofiltration, and biological activated carbon advanced water treatment processes.

Reverse Osmosis

Injectate modeled for treatment by the reverse osmosis process featured a pH of approximately 7.8 after adjustment with lime ($CaOH_2$), dilute TDS (46 mg/L), and, correspondingly, a low ionic strength (0.0015). Cations and anions in the influent were reduced following the treatment process, resulting in concentrations of cations like potassium (less than 1 mg/L), magnesium (less than 1 mg/L), and sodium (less than 10 mg/L) falling below their method detection limits (MDLs), with similarly lower concentrations of anions like phosphate (0.01 mg/L), chloride (less than 10 mg/L), and sulfate (less than 1 mg/L). At York River Treatment Plant, RO-treated waste displayed a calcium-bicarbonate water type. Metals such as iron, manganese, and arsenic exhibited low concentrations, and all near MDLs. RO has limited effect on DO concentrations in the injectate, which ranged around 5 mg/L (HRSD, 2016b).

Nanofiltration

Injectate derived from the nanofiltration process exhibited a pH of approximately 7.8, moderate TDS (262 mg/L), and corresponding ionic strength (0.005). Cations and anions are reduced following the treatment process, but unlike reverse osmosis displayed measurable concentrations of major cations like potassium (7.7 mg/L), magnesium (2.5 mg/L), and sodium (58 mg/L), and anions including phosphate (0.03 mg/L), chloride (125 mg/L, and sulfate (1.8 mg/L). Nanofiltration injectate displayed sodium-chloride chemistry. Concentrations of metals including iron, manganese, and arsenic fell below MDLs. NF injectate also displayed near-saturated concentrations of DO around 5 mg/L.

Biologically Activated Carbon

Injectate originating from biologically activated carbon treatment displayed a pH of approximately 7.8, slightly brackish TDS (615 mg/L), and corresponding ionic strength (0.009). Cations and anions exhibited less reduction following the treatment process compared with nanofiltration and reverse osmosis. Cation concentrations of potassium (13 mg/L), magnesium (10 mg/L), and sodium (103 mg/L) exceeded the concentrations yielded by the membrane (reverse and nanofiltration) treatments. Concentrations of anions like phosphate (0.5 mg/L), chloride (212 mg/L), and sulfate (44 mg/L) also appeared correspondingly higher. Unlike reverse osmosis and nanofiltration, DO concentrations fell to 1.3 mg/L after treatment using biologically activated carbon. Biologically activated carbon injectate featured sodium-chloride water chemistry. Iron and arsenic displayed concentrations higher than the membrane treatment options at 0.002 and 0.73 mg/L, respectively. Iron concentrations

above 0.1 mg/L create significantly large amounts of TSS that can quickly clog an
injection well.

NATIVE GROUNDWATER

Native groundwater quality from the UPA, MPA, and LPA zones was obtained from
nested observation wells maintained by the USGS and NWIS, located 5 miles west
of the York River WWTP. At this location, observation wells installed were screened
in the UPA from 527 to 537 ft below grade (fbg), in the MPA from 820 to 830 fbg,
and in the LPA from 1,205 to 1,215 fbg.

Upper Potomac Aquifer Zone

The UPA featured a slightly alkaline pH (8.2), brackish TSD (1,280 mg/L0), anoxic
water (DO less than 1.0 mg/L), displaying an ionic strength of 0.02. The groundwater
exhibited low amounts of nutrients with concentrations of ammonia, nitrates, and
phosphate falling below 0.01 mg/L. Chloride concentrations approached 500 mg/L,
while sodium concentrations appeared similarly elevated (516 mg/L). Concentrations
of other cations including calcium (4.5 mg/L), and potassium (13.5 mg/L) were comparatively low. Iron concentrations fell around its MDLs (0.01–0.04 mg/L), while
manganese concentrations approached the drinking water MCL of 0.05 mg/L. Water
from the UPA displayed a sodium-chloride chemistry.

Middle Potomac Aquifer Zone

The MPA also featured a slightly alkaline pH (8.0), brackish TDS (2,780 mg/L),
anoxic groundwater (DO less than 1.0 mg/L), with an ionic strength of 0.04. Similar
to the UPA, concentrations of nutrients fell below 0.01 mg/L. Concentrations
of anions, comprising chloride (1,200 mg/L), alkalinity (370 mg/L), and sulfate
73 mg/L), exceeded concentrations encountered in the UPA. Sodium concentrations
appeared similarity elevated at 870 mg/L, while concentrations of other cations,
such as calcium, magnesium, and potassium fell below 15 mg/L. Iron concentrations
were near MDLs (0.01– 0.04 mg/L), while manganese concentrations appeared at
0.02 mg/L. Similar to the UPA, groundwater in the MPA exhibited sodium-chloride
chemistry.

Lower Potomac Aquifer Zone

The Lower Potomac Aquifer displayed a circum-neutral pH (7.7) brackish TDS
(4,580 mg/L), and anoxic groundwater (DO less than 1.0 mg/L). Similar to the other
PAS aquifers, concentrations of nutrients fell below 0.01 mg/L, although phosphate
concentrations, at 0.5 mg/L, appeared notably higher than in the other PAS aquifers.
Concentration of anions, comprising chloride (2,950 mg/L) and sulfate (146 mg/L),
exceeded concentrations encountered in the UPA and MPA. Sodium concentrations were similarly elevated (1,700 mg/L). Concentrations of other cations such
as potassium (25 mg/L) and calcium (51 mg/L) increased over the concentrations
encountered in the other PAS zones. Yet, iron and manganese in the LPA mimicked
groundwater from the other aquifers with concentrations at MDLs and 0.02 mg/L,

respectively. Similar to the other PAS aquifers, groundwater from the LPA displayed sodium-chloride chemistry.

Geochemical Assessment of Injectate and Groundwater Chemistry

In this section, the discussion is about the chemical assessment of the injectate (i.e., the effluent from the advanced water treatment processes, reverse osmosis, nanofiltration, and biologically activated carbon) that potentially is to be injected as injectate into native groundwater. Obviously, as pointed out with the example of the beakers of pure water and sea water mixing that adulterated the pure water into something not potable, HRSD does not want a similar outcome with its injection of its treated wastewater into the Potomac Aquifer's native groundwater. The goal is to determine the appropriate injectate. Keeping the desired outcome and goal in mind, the evaluation of the chemistry of the injectate water and native groundwater from the Potomac Aquifer System revealed the following:

- Reverse osmosis and nanofiltration displayed ionic strengths differing by over one order of magnitude from the native groundwater in the Potomac Aquifer System.
- Biological activated carbon displayed ionic strengths within the same order of magnitude as the Potomac Aquifer System.
- Reverse osmosis exhibits a differing cationic chemistry than the groundwater from the Potomac Aquifer System.

Influence of Ionic Strength

The ionic strength of reverse osmosis diluted, treated water appeared lower than groundwater in the three Potomac Aquifer System zones by at least one order of magnitude. By comparison, the ionic strength displayed by nanofiltration differed from the Lower Potomac Aquifer by over nine orders of magnitude. The ionic strength of biologically treated carbon, although lower than the Potomac Aquifer System aquifers, fell within the same order of magnitude. The low ionic strength of reverse osmosis compared to the Potomac Aquifer System groundwater represents a concern for injection operations, particularly for reverse osmosis' potential to disperse clay minerals. Clay dispersion represents an electro-kinetic process (Meade, 1964; Reed, 1972; and Gray and Rex, 1966), where an electrostatic attraction between negatively charged clay particles is opposed by the tendency of ions to diffuse uniformly throughout an aqueous solution. One of the most important factors leading to the dispersion of clay minerals involves a change in the double-layer thickness of a clay particle. A double layer of ions lies adjacent to the clay mineral surface or between the mineral's structural layers because a negative charge attracts cations toward the surface. As the fluid must maintain electrical neutrality, a more diffuse layer of anions surrounds the cations.

As in brackish water, when the concentration of ions is large, the double layer around the particle or between the clay's structure layers gets compressed to a smaller thickness. Compressing the double layer causes particles to coalesce, forming larger aggregates. This process is called clay flocculation. When the ionic concentration of a fluid invading the aquifer is significantly lower than the native groundwater, the diffuse-double layer expands, forcing clay particles and the structural layers within clay minerals apart. The expansion prevents the clay particles from moving closer together and forming an aggregate. The tendency toward dispersion is measured in clay minerals by their zeta potential—that is, where colloids with high zeta potential are electrically stabilized while colloids with low zeta potentials tend to coagulate or flocculate—according to the following relationship:

$$Z = 4\pi\delta q/D \qquad (12.18)$$

where

   Z = zeta potential
   δ = thickness of the zone of influence surrounding the charge particle
   q = charge on the clay particle before attaching cations
   D = dielectric constant of the liquid

For any solution and clay mineral, reducing the zeta potential involves lowering the thickness of the zone of influence. Substituting small, double- or triple-charged cations such as Ca^{+2} or Al^{+3}, respectively, in place of large singly charged and hydrated ions like Na^{+2}, lowers the zeta potential, permitting clay particles to coalesce. This behavior explains the tendency for sodium to cause clay dispersion, while calcium and aluminum induce its flocculation.

Aquifer sands containing more complex clay minerals like mixed-layer clays and smectite group clays that display small particle sizes, yet large surface charges usually exhibit the greatest sensitivity to fresh water (Brown and Silvey, 1977). As little as 0.4%t smectite in a sand body has reduced the aquifer's hydraulic conductivity by 55% after exposure to fresher water (Hewitt, 1963). Mixed-layer clays and smectite encountered in cores from the UPA and MPA at the city of Chesapeake exhibited abundance equaling trace (less than 1–4% of the whole rock composition of the sand).

In the 1970s, the USGS tested an aquifer storage and recovery facility in Norfolk, Virginia; it exhibited greater than 50% reduction in injectivity after only 150 minutes of starting injection operations (Brown and Silvey, 1977). The aquifer storage and recovery well were installed in the UPA, screening nearly 85 ft of sand in the unit. The USGS employed nuclear, electrical, and mechanical geophysical logging techniques to evaluate the origin of the injectivity losses and discriminate between the causes of physical plugging documented at other sites, like TSS loading. Injectivity losses caused by physical plugging from TSS loading typically occur at discrete zones through the well screen. In contrast, geophysical logging of the aquatic storage and recovery test well at Norfolk showed hydraulic conductivity losses distributed evenly across the entire screen. Also, in comparison to physical plugging by TSS, which responds positively to mechanical and chemical rehabilitation techniques, the USGS was unable to restore even a fraction of the well's original injectivity.

USGS used calcium chloride ($CaCl_2$ greater than 10,000 mg/L) solution to treat the wellbore and proximal aquifer to arrest the declining injectivity. The double charged calcium cation forms a stronger particle and inter-layer bond than the mon-ovalent cation, sodium. Using a concentrated solution ensures calcium exchanges for sodium at the maximum number of sites. After applying the treatment at Norfolk, the injectivity of the aquifer storage and recharge test well remained stable over two more test cycles before the project ended. Concentrated solutions containing the trivalent aluminum proved to be effective in stabilizing clay minerals prone to dis-persion in the presence of dilute injectate (Civan, 2000). Applying a calcium or alu-minum chloride treatment to the Potomac Aquifer System before initiating injection operations, offers a viable alternative for stabilizing clay minerals in situ, precluding formation damage, and injectivity loss should regulators select reverse osmosis as the most viable method for protecting local water users. These treatments could also benefit injection operations using nanofiltration or biologically activated carbon as the preferred injectate.

Cation Exchange

In addition to differing ionic strengths, reverse osmosis, as calcium carbonate water, differs from the sodium-chloride chemistry encountered in the UPA, MPA, and LPA. As previously described, the doubly charged calcium ion should benefit the long-term stability of clay minerals where calcium exchanges for sodium. However, calcium exhibits a large ionic radius that can damage clay mineral when entering the position left by the sodium, fragmenting the edges of the mineral, and mobilizing the fragments in the aquifer environment.

Iron and Manganese

None of the injectates or native groundwater from the Potomac Aquifer Systems aquifers appears to exhibit problematic concentrations of iron or manganese. Iron concentrations in reverse osmosis and nanofiltration effluent typically occurred below MDLs. During injection operation, iron and manganese contained in the injectate or native groundwater can precipitate oxide and hydroxide minerals when exposed to DO. Formation of these minerals presents a problem if they precipitate close to the well bore, which is a zone featuring small surface areas sensitive to physical plugging. Accordingly, the absence of iron and manganese in injectate or native groundwater benefits injection operations.

LITHOLOGY OF THE POTOMAC AQUIFER SYSTEM

The lithology and minerals comprising the Potomac Aquifer System aquifers is described in this section, starting with the general composition across the study area, and then focusing on cores collected from the UPA and MPA near the city of Chesapeake, Virginia.

LITHOLOGY

As previously mentioned, the Potomac Aquifer System consists of three-discrete aquifer zones (Upper Potomac Aquifer, Middle Potomac Aquifer, and Lower

Potomac Aquifer) named for their position in the section. Deposited in river (fluvial) and shallow marine environments, the aquifers consist of coarse to fine sands with occasional gravel, interbedded with thin gray to pale green clays (Treifke, 1973). The aquifers are separated by clay beds of thicknesses exceeding 20 ft. However, thinner clay beds transect the sand units in the Middle Potomac Aquifer and the Lower Potomac Aquifer. Because of the abundance of clay beds, the Middle Potomac and Lower Potomac often consist of multiple, stacked units requiring repeated screen and blank combinations for supply wells installed in these aquifers.

Sands are comprised primarily of quartz (Mend and Harsh, 1988), often reaching amounts exceeding 90% by weight, forming the predominant framework mineral. Accessory minerals include orthoclase, muscovite, glauconite, and locally, lignite. Trace minerals mostly occupy the interstitial spaces in the sands and comprise biotite, pyrite, siderite, magnetite, and clays.

CITY OF CHESAPEAKE AQUIFER STORAGE AND RECOVERY FACILITY CORE SAMPLES

In 1989, at the city of Chesapeake's aquifer storage and recovery facility, 10 core samples were collected from the Upper Potomac Aquifer and Middle Potomac Aquifer, at depths ranging from 560 to 835 fbg. The cores were submitted to Mineralogy Inc., a laboratory specializing in mineralogical assays, for the following analyses:

- Specific gravity.
- Porosity.
- Permeability.
- X-ray diffraction.
- Cation exchange capacity (CEC).
- Grain size distribution.
- Energy dispersive chemical analysis.
- Scanning electron microscopy.

Potomac Aquifer System sediments found at the city of Chesapeake locations, even though they are located approximately 35 miles apart, should display similar characteristics as those underlying the Yorktown Wastewater Treatment Plant. Because aquifer characteristics like grain size, sorting, textural maturity, porosity, and permeability decline moving downdip, Potomac Aquifer System sediments at Yorktown Wastewater Treatment Plant should display characteristics better suited to injection operations than at the city of Chesapeake.

Core samples from the Upper Potomac Aquifer and Middle Potomac Aquifer were composed of coarse to very coarse-grained sands, in a medium-grained matrix. Aquifer sands appeared conglomeratic and unsorted. However, as unconsolidated sands they displayed open pore spaces yielding good porosity (21–34%) and air permeability. Grain size diminished with depth. Samples from the deeper portions of the Middle Potomac Aquifer exhibited a medium-grain size with a larger percentage of fine sands than shallower samples. Most clay minerals were found in interstitial spaces of the aquifers and showed a high degree of crystallinity, which suggests they formed after deposition and burial (i.e., were authigenic).

Sands from the Upper Potomac Aquifer and Middle Potomac Aquifer consisted of 84–89% quartz with 8–12% potassium and plagioclase feldspar, classifying the sands as subarkosic, or lithic arkosic. Trace (less than 10%) amounts of calcite and dolomite were detected in every sample of the aquifer sands. Clay minerals, comprising kaolinite, illite/mica, and smectite made up 4% of the same samples. The iron carbonate mineral, siderite ($FeCO_3$) was encountered in a confining bed sample (595 fbg) at an amount up to 19%. Siderite was also encountered in an aquifer core (685.2 fbg), at trace amounts.

Sands from the Upper Potomac Aquifer and Middle Potomac Aquifer consisted of 84–89% quartz with 8–12% potassium and plagioclase feldspar, classifying the sands as subarkosic, or lithic arkosic (i.e., a type of sandstone containing at least 25% feldspar). Trace (less than 10%) amounts of calcite and dolomite were detected in every sample of the aquifer sands. Clay minerals, comprising kaolinite, illite/mica, and smectite made up 4% of the same samples. The iron carbonate mineral, siderite ($FeCO_3$) was encountered in a confined bed sample (595 fbg) at amounts up to 19%. Siderite was also encountered in an aquifer core (685.2 fbg) at amounts up to 19%. Siderite was also encountered in an aquifer core (685.2 fbg), at trace amounts.

Permeabilities in air (intrinsic permeability) ranged from 1280 to 5900 millidarcies. Generally, intrinsic permeability and porosity values declined with depth, so the greatest permeabilities were encountered in cores from the Upper Potomac Aquifer. Intrinsic permeability displayed minimal anisotropy with horizontal and vertical values from the same core yielding near equal permeabilities. Cation exchange capacity or CEC refers to the number of exchangeable cations per dry weight that a soil can hold, at a given pH, and are available for exchange with the soil-water solution which is influenced by the amount and type of clay and the amount of organic matter (Drever, 1982). CEC serves as a measure of soil fertility, nutrient retention capacity, and the capacity to protect groundwater from cation contamination. The CEC of minerals contained in confining beds often control the cation chemistry in the adjacent aquifers by exchanging cations across the contact between the units. For injection purposes, knowing the CEC of the aquifer and confining bed materials can help assess how these materials will react with recharge water displaying a specific cation ionic chemistry. CEC is expressed as milliequivalents of hydrogen per 100 grams (g) of dry soil (meq+/100g).

All ten samples at the Chesapeake site were analyzed for CEC. Sodium represented the most dominant exchangeable cation followed by magnesium, calcium, and potassium. The confining bed sample at 685.2 fbg, displayed the most elevated CEC at 12.5 meq/100g of core. Aquifer sand samples from the Upper Potomac Aquifer and Middle Potomac Aquifer exhibited CECs for sodium ranging from 0.7 to 3.9 meq/100 g of core. Sodium, a monovalent ion in the exchange position of clays, will not benefit from injection operations.

Despite the dominance of sodium, CEC values from cores from the city of Chesapeake were low, suggesting that the clays should display a minimal tendency to exchange cations. In environments showing more elevated CECs, divalent ions like calcium or magnesium in the injectate can exchange with sodium, temporarily disrupting the clay's atomic structure. Over the long-term, replacing sodium with a divalent ion will strengthen the clay mineral's atomic structure eventually transitioning to a stable smectite.

MINERALOGY—GEOCHEMICAL MODELING

The thermodynamic equilibrium model PHREEQC (Parkhurst, 1995) was used to gain a greater understanding of the stability of the clay minerals in the Potomac Aquifer System aquifers beneath the Yorktown Wastewater Treatment Plant, based on the native groundwater and injectate chemistries. As previously described, the stability of clay minerals can control the success of injection operations in sandy aquifers like the Potomac Aquifer System.

Thermodynamic equilibrium models consist of computer programs using a relatively sophisticated set of equations (Davies, 1962; Truesdell and Jones, 1974; Debye and Huykel, 1923) to stimulate the chemical equilibrium of a solution under natural or laboratory conditions, and to simulate the effects of chemical reactions. These models perform the following types of calculations:

- Correct all equilibrium constants to the temperature of the specific sample.
- Calculates speciation: the distribution of chemical species by element by solving a matrix of equations.
- Calculation's activity coefficients of each chemical species.
- Calculates the state of saturation for potential mineral species that occur in equilibrium with the samples water chemistry. These calculations identify potential mineral species, whether they will dissolve or precipitate under the changing conditions consistent with Aquifer Storage and Recovery operations.
- The models perform a wide variety of calculations related to oxidation-reduction processes.

Thermodynamic equilibrium computer models represent a power tool for predicting chemical behavior in a natural system. Manual manipulation of the same equations performed by these programs are time consuming and prone to calculation errors.

Stability of Clay Minerals

PHREEQC was employed for evaluating the stability of clays contained in the $CaO\text{-}Al_2O_3\text{-}SiO_2\text{-}H_2O$; $NaO\text{-}Al_2O_3\text{-}SiO2\text{-}H^2O$; and $K_2O\text{-}Al_2O_3\text{-}SiO_2\text{-}H_2O$ mineral systems. Minerals contained in these systems represent clays and their weathering products (gibbsite, kaolinite) commonly found in sediments of the Potomac Aquifer System. The simulations' objective involved determining how native groundwater chemistries fall into the stability fields of clay minerals and identifying potential instabilities. Along with ambient clay stabilities, PHREEQC simulates how clay can evolve during the exchange of cations. However, the program does not address instability arising from introducing injectate of a differing ionic strength.

The chemistry of groundwater from the Upper Potomac Aquifer, Middle Potomac Aquifer, and Lower Potomac Aquifer, along with potential injectate waters from the Yorktown Treatment Plant was plotted on stability diagrams for three systems describing common clay minerals ($CaO\text{-}Al_2O_3\text{-}SiO_2\text{-}H_2O$; $NaO\text{-}Al_2O_3\text{-}SiO2\text{-}H^2O$; and $K_2O\text{-}Al_2O_3\text{-}SiO_2\text{-}H_2O$). Common clay minerals including smectite, beidellite, montmorillonite, illite, and the gibbsite (weathering product) were over saturated in recharge water samples, which suggests a tendency to precipitate over time.

When injecting waters of incompatible ionic strength or differing cations, damage to clay minerals can arise; however, this was not a concern during injection operations in the Potomac Aquifer System because precipitation of clay minerals represented a relatively minimal matter regarding permeability loss. Moreover, the precipitation of clay minerals requires significant amounts of geologic time, rather than the relatively short service life of an injection facility.

Simulated Injectate–Water Interactions

Along with characterizing clay mineral stability during injection operations, PHREEQC was also employed in assessing reactions originating from mixing the three injectate types, reverse osmosis, nanofiltration, and biologically activated carbon injectates, with native groundwater from each of the PAS' aquifers, and reactions between the injectate and elected reactive minerals in the PAS aquifers. Mixing reactions occur when injectate interfaces with native groundwater. As injection operations proceed, injectate drives the mixing interface further into the PAS aquifers.

Surface areas in an aquifer undergoing injection are small around the injection wellbore but increase geometrically with distance away from the well. The larger surface areas away from the injection wells help buffer reactions that cause permeability losses. Thus, these reactions diminish in importance as injection operations progress. Reactions between injectate and reactive minerals cause the following concerns:

- Permeability losses with the precipitation of iron or manganese oxide minerals.
- Leaching of environmentally problematic constituents like iron, manganese, and arsenic along with other metals, depending on the ambient mineralogy.

Mixing

Mixing could arise during injection operations, including:

- Mixing between the injectate and native groundwater.
- Mixing between groundwater from the Upper Potomac Aquifer, Middle Potomac Aquifer, and Lower Potomac Aquifer in the injection wellbore.

Mixing Injectate and Native Groundwater

As mentioned, mixing reactions prove most troublesome around the injection well-bore. One common reaction involves the precipitation of oxide minerals when injectate-containing dissolved oxygen contacts dissolved iron or manganese entrained in the injector or native groundwater. Although each injectate from the advanced water treatment processes contained measurable concentrations of dissolved oxygen, dissolved iron, and manganese concentrations, (with the exception of biologically activated carbon) were absent in the injectate and native groundwater. Despite the absence of iron and manages other minerals can also precipitate during mixing.

PHREEQC modeling was employed to simulate mixing between the injectate and native groundwater chemistries at a 1:1 ratio in order to evaluate the mixing between the differing water types. The modeling was also used to evaluate reactions between the native groundwater chemistries in the Upper Potomac Aquifer, Middle Potomac Aquifer, and Lower Potomac Aquifers as they are mixed in an injection wellbore, simulating an injection well screening the three Potomac Aquifer System aquifers.

During the mixing simulations, important reactions were tracked including the potential precipitation of metal oxide, hydroxide, sulfate, and carbonate minerals along with dissolution of silicates, including clays. Because of the similar bulk chemistry of the three injectates and native groundwater from the UPA, MPA, and LPA, potential mineral suites identified by PHREEQC in the mixed water and their saturation indices were repeated across the nine mixtures. The *saturation index* (SI) of a mineral determines whether the mineral occurs in equilibrium (Langmuir, 1997) with mixed water chemistry (SI equal to 0.0); is under saturated, and if present, should dissolve (SI less than 0.0); or is supersaturated (SI greater than 0.0) and should precipitate. Estimation of saturation indices are usually not exact, often varying over ±0.3 units, depending on the composition of the solution. The SI provides a guideline on how minerals will behave in a water sample.

Of the common minerals and their weathering products identified in nine combinations of mixed water quartz, the most common mineral in sands of the Potomac Aquifer System returned a slightly oversaturated SI (SI equal to 0.66–0.87) for all simulations. Oversaturation of quartz and near-equilibrium saturation indices for less crystalline forms of silica such as chalcedony (SI equal to −0.06 to 0.29) and cristobalite (SI equal to 0.02–0.37) indicates feldspars are dissolving, releasing silica. Gibbsite ($Al(OH)_3$) appeared oversaturated in the mixed water consistent with the dissolution of feldspar, and precipitation of residual byproducts in a weathered environment.

Other minerals potentially reacting in the mixed water included the carbonates, comprising calcite, aragonite, and siderite. Calcite ($CaCo_3$) appeared uniformly undersaturated (SI equal to −1.42 to −0.24) in all the mixed waters, suggesting it will not precipitate. Calcite, in an undersaturated state, benefits injection operations by not precipitating, blocking pore spaces, and reducing the permeability of the aquifer. Aragonite, an isomorph of calcite, shows similar indices, ranging from −1.56 to −0.39. The iron carbonate mineral ($FeCO_3$), siderite displayed SI values similar to calcite and aragonite with strongly undersaturated indices ranging from −8.09 to −11.73.

Other important minerals included gypsum and jarosite, a weathering product of iron-bearing mineral sand sulfides. Similar to the carbonates, with one exception, gypsum and jarosite exhibited undersaturated indices ranging from −2.21 to −3.37 and −5.62 to 0.33, respectively. As a single exception, biologically activated carbon mixing with groundwater from the Lower Potomac Aquifer resulted in near-equilibrium SI for jarosite.

Mixing in the Injection Wellbore

With an open conduit extending between the three Potomac Aquifer System aquifers at York River Treatment Plant, groundwater will mix in the injection wellbore before the start of injection operations. Once injection operations start, injectate will displace the groundwater away from the wellbore so mixing groundwater will no longer present an issue.

An examination of static water-level elevations at the nested NWIS wells near the York River Wastewater Treatment suggests a vertically downward hydraulic gradient of 0.085 feet/foot (ft/ft) occurring between Upper Potomac Aquifer and Middle Potomac Aquifer, while an upward gradient of 0.031 ft/ft appears between the Lower Potomac Aquifer and Middle Potomac Aquifer. The differing gradient directions impose converging flow in the wellbore. Accordingly, groundwater will flow in through intervals screening the Upper Potomac Aquifer and Lower Potomac Aquifer and out through the screen against the Middle Potomac Aquifer, promoting the mixing of the three water types in the wellbore, rather than stratification.

In the geochemical simulation, groundwater from the Upper Potomac, Middle Potomac Aquifer, and Lower Potomac Aquifer was mixed at even proportions between the three units. To maintain a conservative approach to the simulations, ferrous iron (Fe II) concentrations at Upper Potomac Aquifer, Middle Potomac Aquifer, and Lower Potomac Aquifer were assumed to occur at 0.1 mg/L, the MDL for iron. Concentrations in the mixed water oxidized form Fe II to ferric iron (Fe III), remain at a concentration of 0.1 mg/L.

No deleterious reactions associated with mixing in the Aquifer Storage and Recovery wellbore was detected through modeling. The pH of the mixed water declined slightly to 7.83. SI for calcite and siderite appeared near-equilibrium at −0.08 and −0.04, respectively. This suggests that these minerals should neither dissolve nor precipitate in the mixed water. The mixed water displayed sodium-chloride chemistry similar to groundwater from the three Potomac Aquifer System aquifers. Aragonite, gypsum, and jarosite displayed unsaturated SIs indicating that the mineral should not precipitate in the mixed water. The weathering products of clay minerals, gibbsite remained saturated roughly the same as the individual groundwater chemistries.

INJECTATE AND AQUIFER MINERAL REACTIONS

In addition to reactions between dilute injectate and clay minerals, dissolved oxygen in the injectate can react with reduced metal-bearing minerals releasing metals and other constituents, which can compromise the quality of water disposed in the

Potomac Aquifer System aquifers. These reactions should not affect injection operations, but can result in environmental concerns, prompting the attention of regulators.

Analysis of cores from the Aquifer Storage and Recovery project at the city of Chesapeake encountered microcrystalline siderite in the interstices of aquifer sands and as larger crystalline forms in adjoining confining beds.

Pyrite is another reduced, metal-bearing mineral common to the Virginia Coastal Plain Aquifers, including the Potomac Aquifer System (Meng and Harsh, 1988; McFarland and Bruce, 2006). Although not detected in core samples from the city of Chesapeake, pyrite and siderite typically occur together in sediments subject to flooding by marine and freshwater systems (Postma, 1982), typical of the near coastal environment in which the formation bearing the Potomac Aquifer System aquifers was deposited. Accordingly, pyrite was considered in the geochemical modeling evaluation.

The primary metal in siderite and pyrite is FE II but both minerals can also contain cadmium and manganese. Additionally, pyrite occasionally contains varying amounts of arsenic (Evangelou, 1995). Dissolved oxygen in the three injectates should range between 1.3 and 5 mg/l, providing a source for oxidizing reactions.

Siderite Dissolution

Dissolved oxygen reacts with siderite to release Fe II and CO_3^{2-} (carbon trioxide). Upon encountering dissolved oxygen, Fe II oxidizes to Fe III, which acts as a strong oxidant, continuing the dissolution of siderite. At equilibrium, a small amount of siderite can release large amounts of Fe II into the surrounding pore water.

PHREEQC was employed to simulate potential reactions between the injectates and siderite. In this reaction, one mole of siderite was reacted with reverse osmosis, nanofiltration, and biologically activated carbon, each containing dissolved oxygen concentrations ranging from 1 to 7 mg/L in 1.0 mg/L increments. Resulting Fe II at dissolved concentrations of 7 mg/L ranged from 90 to 130 mg/L for reverse osmosis and biologically activated carbon, respectively, with nanofiltration exhibiting concentrations between reverse osmosis and biologically activated carbon. Even acting with only 1 mg/L dissolved with siderite produce Fe II concentrations ranging from 50 to 90 mg/L for reverse osmosis and biological activated carbon, respectively.

Bicarbonate concentrations increased from 95 mg/L for reverse osmosis at 1 mg/L dissolved oxygen to over 200 mg/L at 7 mg/L dissolved oxygen for biologically activated carbon. A portion of the total bicarbonate comprised carbonic acid, lowering the simulated pH of the pore water from 7.8 to 6.91 for reverse osmosis at 7 mg/L dissolved oxygen. The pH of reverse osmosis, nanofiltration, and biologically activated carbon dropped below 7.15 when reacting siderite with a dissolved oxygen of only 1 mg/L reverse osmosis, the most dilute injectate, exhibited the lowest capacity to buffer the pH during the reaction between siderite and dissolved oxygen.

Pyrite Oxidation

Although pyrite was not encountered in the cores collected at the city of Chesapeake, its appearance elsewhere in the Potomac Aquifer System, and its deleterious reactions when encountering dissolved oxygen make evaluating the mineral an important part of any injection feasibility study. Reacting dissolved oxygen with pyrite releases Fe II and the bisulfide ion (Evangelou, 1995). Upon encountering dissolved oxygen, Fe II oxidized to Fe III, which also acts as a strong oxidant, continuing the oxidation of pyrite. The bisulfide ion (S_2^{2-}) further reacts with dissolved oxygen to form sulfuric acid (H_2SO_4), lowering the pH of the surrounding pore water.

PHREEQC was used to simulate an operational injection scenario to predict the chemistry of effluent containing varying amounts of dissolved oxygen, exposed to pyrite in the Potomac Aquifer System aquifer matrices. Pyrite was equilibrated with reverse osmosis, nanofiltration, and biologically activated carbon containing concentrations ranging from 1.0 to 7.0 mg/L in 1.0 mg/L increments.

Similar to the siderite simulations, the injectate chemistries were equilibrated with one mole of pyrite. Thus, simulations provide conservative results, overestimating the concentrations of iron, sulfate, and arsenic in the recovered water. Where present, in the Atlantic Coastal Plain aquifers, pyrite comprises less than 1% of the whole rock composition or 0.05 to 0.1 moles.

Modeling results showed that Fe II concentrations increased from 3 to over 10 mg/L, while sulfate increased at twice this rate. Sulfate concentrations simulated with biologically activated carbon were elevated above the other effluent types by its initial concentration of 44 mg/L. Similar to the simulations with siderite, reverse osmosis exhibited the greatest decline in pH after reacting with pyrite, with the pH declining from 7.8 to less than 6.8.

The iron concentrations simulated from the modeling are considered conservative as a large portion of Fe II released by the oxidation of pyrite will precipitate as hydrous ferric oxides (HFO). The HFO typically precipitates on the pyrite mineral surface, progressively reducing its reactivity (passivate). Moreover, these surfaces can adsorb Fe II migrating in the aquifer environment. In the absence of pyrite, these simulations illustrate potential groundwater quality problems that can emerge from effluent containing dissolved oxygen in the presence of this reactive mineral.

Arsenic

The release of arsenic from pyrite was simulated with PHREEQC. As a substitution for sulfur, arsenic concentrations were estimated at 1% by weight of the mass of pyrite. Similar to previous simulations, one mole of pyrite was equilibrated with reverse osmosis, nanofiltration, and biologically activated carbon containing dissolved oxygen concentrations ranging from 1 to 7 mg/L. Applying this approach, arsenic concentrations increased from 31 to nearly 58 micrograms per liter (µg/L) in reactions with reverse osmosis, and from 58 to 85 µg/L during reactions with biologically activated carbon.

Similar to relationships between pyrite and iron, arsenic concentrations simulated with PHREEQC, represent conservative conditions. HFO surfaces in aquifer settings display strong affinity for adsorbing the oxyanions of arsenic and lowering its concentrations in groundwater. Moreover, continuing oxidation of pyrite precipitates HFO on the mineral surface, passivating the mineral, also diminishing the concentration of constituent released during oxidation reactions.

MITIGATING PYRITE OXIDATION

At Aquifer Storage and Recovery facilities recharging beneath the Atlantic Coastal Plain, siderite and pyrite dissolution is addressed by increasing the pH of the injectate by adding potassium and sodium hydroxide (caustic). Increasing the injectate pH raises it above the solubility limit of iron, buffering the dissolution of iron-bearing minerals. Hydroxyl ions in sodium and potassium hydroxide will react with Fe II released from siderite or pyrite, oxidizing Fe II and Fe III. It precipitates HFO on the surface of these minerals, which then passivates the minerals to future reactions in the aquifer environment. In addition to isolating the minerals, HFO surfaces display excellent adsorption properties, adsorbing metals migrating in the aquifer environment including arsenic and Fe II. Iron adsorbs as a surface precipitate on HFO, while these surfaces exhibit an affinity for adsorbing arsenic at pH values encountered in groundwater environments (Dzomback and Morel, 1990).

PHREEQC was employed to simulate adjusting the pH of reverse osmosis, nanofiltration, and biologically activated carbon injectate-containing dissolved oxygen from 7.8 to 8.5 in the presence of pyrite. Consistent with other simulations, dissolved oxygen was varied between 1.0 and 7.0 mg/L. Fe II concentrations approaching 10 mg/L during reactions with only dissolved oxygen, fell to less than 1.0×10^{-7} mg/L in simulations with injectate pH adjusted to 8.5. Fe II was nearly completely oxidized to Fe III, which precipitated as $Fe(OH)_3$. The modeling results illustrate how well adjusting the pH of the injectate can control Fe II concentrations. As equilibrium simulations, the modeling did not account for the reactions passivating siderite and pyrite over time, which also reduces Fe II concentrations in groundwater (HRSD, 2016b).

SUSTAINABLE WATER INITIATIVE FOR TOMORROW (SWIFT)

Figure 12.4 presents a process flow block diagram of 8-unit processes (8-step process) connected in a treatment train for the SWIFT Demonstration Facility at HRSD's Nansemond Treatment Plant in Suffolk, Virginia. In addition, a very brief fundamental explanation of each of the 8-unit processes is provided below.

SWIFT UNIT PROCESS DESCRIPTION

In this section each of the SWIFT unit processes are described.

> STEP 1: Influent pump station (see Figure 12.5). Wastewater effluent from the Nansemond Treatment Plant is directed to the SWIFT processing site on the plant location. Thus, the highly treated water from the Nansemond Treatment Plant is pumped to the Research Center's advanced treatment facility (SWIFT) where it undergoes advanced treatment within the unit processes housed within; the highly treated SWIFT effluent (of better drinking water quality than that contained within the Potomac Aquifer) is out-falled (injected) into the Upper, Middle, and Lower Potomac Aquifer layers.

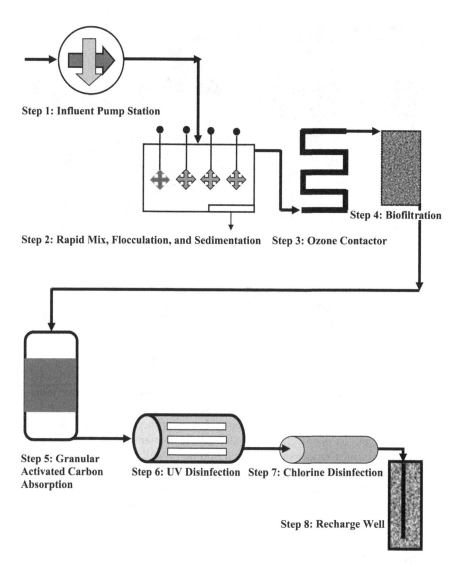

Step 1: Influent Pump Station

Step 4: Biofiltration

Step 2: Rapid Mix, Flocculation, and Sedimentation   Step 3: Ozone Contactor

Step 5: Granular Activated Carbon Absorption

Step 6: UV Disinfection   Step 7: Chlorine Disinfection

Step 8: Recharge Well

FIGURE 12.4   Process flow diagram for SWIFT demonstration facility.

STEP 2: Mixing, flocculation, and sedimentation—This unit process removes suspended solids by settling large particles to the bottom of the water column.

STEP 3: Ozone Contact—This unit process breaks down organic material and provides disinfection.

STEP 4: Biologically Active Filtration—This unit process filters out suspended particles, pathogens, and removes dissolved organic compounds through microbiological activity.

FIGURE 12.5   SWIFT influent pump station. Photo by F. R. Spellman.

STEP 5: Granulated Activated Carbon Contactors—This unit removes trace organic compounds and prepares the water tor ultraviolet disinfection.

STEP 6: Ultraviolet (UV) Disinfection—This unit process provides a barrier to pathogens by disinfecting the water with high intensity ultraviolet light.

STEP 7: Chlorine Contact and Chemical Addition—This unit process provides disinfection of finished water using chlorine and serves as an additional barrier to pathogens. Chemical addition is used on the disinfected water and is adjusted by small doses to match the geochemistry of the water already in the aquifer (STEP 8).

THE ULTIMATE BOTTOM LINE

HRSD's SWIFT initiative is a far-thinking innovative initiative that potentially offers enormous benefits not only for the Hampton Roads region but also for any region with similar needs, issues, and/or problems. To conclude with the bottom line of the SWIFT process and its benefits presented in this account, we must recognize the water challenges faced by those in the Hampton Roads region. These water challenges consist of questions that only operational time and operational adjustments can and will eventually answer. These questions are:

- Will SWIFT restore the Chesapeake Bay?
- Will SWIFT mitigate groundwater depletion?
- Will SWIFT prevent saltwater intrusion?

- Will SWIFT counter relative sea-level rise?
- Will SWIFT prevent recurrent flooding?
- Will SWIFT prevent sanitary sewer overflows?
- Will SWIFT be affordable?

The jury is still out on whether these questions will be answered in the positive. We do not know what we do not know at this precise moment. However, early, very early observations, measurements, and results indicate promise.

SOLID-SIDE OF HUMAN WASTE

Raw, untreated human waste is sometimes used beneficially as soil fertilizer, generally in under-developed parts of the globe. In developed parts of the globe a form of night soil, what we call wastewater, the solid side, is used in agriculture. The difference is the use of untreated and treated human waste. Wastewater treatment removes solids from the wastestream before the liquid effluent is discharged to its receiving waters. What remains to be disposed of is a mixture of solids and wastes, called *process residuals*—more commonly referred to as *sludge* or *biosolids*.

Note: Sludge is the commonly accepted name for wastewater solids. However, if wastewater sludge is used for beneficial reuse (e.g., as a soil amendment or fertilizer), it is commonly called *biosolids*.

DID YOU KNOW?

The most costly and complex aspect of wastewater treatment can be the collection, processing, and disposal of sludge. This is the case because the quantity of sludge produced may be as high as 2% of the original volume of wastewater, depending somewhat on the treatment process being used. Because sludge can be as much as 97% water content, and because the cost of disposal will be related to the volume of sludge being processed, one of the primary purposes or goals (along with stabilizing it so it is no longer objectionable or environmentally damaging) of sludge treatment is to separate as much of the water from the solids as possible. Sludge treatment methods may be designed to accomplish both of these purposes.

Note: Sludge treatment methods are generally divided into three major categories: thickening, stabilization, and dewatering. Many of these processes include complex sludge treatment methods (i.e., heat treatment, vacuum filtration, incineration, and others).

SLUDGE/BIOSOLIDS: BACKGROUND INFORMATION

When we speak of *sludge* or *biosolids*, we are speaking of the same substance or material; each is defined as the suspended solids removed from wastewater during sedimentation, and then concentrated for further treatment and disposal or reuse.

The difference between the terms *sludge* and *biosolids* is determined by the way they are managed.

Note: The task of disposing, treating, or reusing wastewater solids is called *sludge* or *biosolids management.*

Sludge is typically seen as wastewater solids that are "disposed" of. Biosolids is the same substance managed for reuse—commonly called beneficial reuse (e.g., for land application as a soil amendment, such as biosolids compost). Note that even as wastewater treatment standards have become more stringent because of increasing environmental regulations, so has the volume of wastewater sludge increased. Also note that before sludge can be disposed of or reused, it requires some form of treatment to reduce its volume, to stabilize it, and to inactivate pathogenic organisms.

Sludge forms initially as a 3–7% suspension of solids, and with each person typically generating about 4 gallons of sludge per week, the total quantity generated each day, week, month, and year is significant. Because of the volume and nature of the material, sludge management is a major factor in the design and operation of all water pollution control plants.

Note: Wastewater solids treatment, handling, and disposal account for more than half of the total costs in a typical secondary wastewater treatment plant.

Sources of Sludge

Wastewater sludge is generated in primary, secondary, and chemical treatment processes. In primary treatment, the solids that float or settle are removed. The floatable material makes up a portion of the solid waste known as scum. Scum is not normally considered sludge; however, it should be disposed of in an environmentally sound way (i.e., until we find a beneficial reuse for it). The settleable material that collects on the bottom of the clarifier is known as *primary sludge*. Primary sludge can also be referred to as raw sludge because it has not undergone decomposition. Raw primary sludge from a typical domestic facility is quite objectionable and has a high percentage of water, two characteristics that make handling difficult.

Solids not removed in the primary clarifier are carried out of the primary unit. These solids are known as *colloidal suspended solids*. The secondary treatment system (i.e., trickling filter, activated sludge, etc.) is designed to change those colloidal solids into settleable solids that can be removed. Once in the settleable form, these solids are removed in the secondary clarifier. The sludge at the bottom of the secondary clarifier is called *secondary sludge*. Secondary sludges are light and fluffy and more difficult to process than primary sludges—in short, secondary sludges do not de-water well.

The addition of chemicals and various organic and inorganic substances prior to sedimentation and clarification may increase the solids capture and reduce the amount of solids lost in the effluent. This *chemical addition* results in the formation of heavier solids, which trap the colloidal solids or convert dissolved solids to settleable solids. The resultant solids are known as *chemical sludges*. As chemical usage increases so does the quantity of sludge that must be handled and disposed of. Chemical sludges can be very difficult to process; they do not de-water well and contain lower percentages of solids.

SLUDGE CHARACTERISTICS

The composition and characteristics of sewage sludge vary widely and can change considerably with time. Notwithstanding these facts, the basic components of wastewater sludge remain the same. The only variations occur in quantity of the various components as the type of sludge and the process from which it originated changes. The main component of all sludges is *water*. Prior to the treatment, most sludge contain 95–99+% water. This high-water content makes sludge handling and processing extremely costly in terms of both money and time. Sludge handling may represent up to 40% of the capital cost and 50% of the operation cost of a treatment plant. As a result, the importance of optimum design for handling and disposal of sludge cannot be overemphasized. The water content of the sludge is present in a number of different forms. Some forms can be removed by several sludge treatment processes, thus allowing the same flexibility in choosing the optimum sludge treatment and disposal method. The forms of water associated with sludges include:

- *Free water*—water that is not attached to sludge solids in any way. This can be removed by simple gravitational settling.
- *Floc water*—water that is trapped within the floc and travels with them. Its removal is possible by mechanical de-watering.
- *Capillary water*—water that adheres to the individual particles and can be squeezed out of shape and compacted.
- *Particle water*—water that is chemically bound to the individual particles and can't be removed without inclination.

From a public health view, the second and probably more important component of sludge is the *solids matter*. Representing from 1 to 8% of the total mixture, these solids are extremely unstable. Wastewater solids can be classified into two categories based on their origin—organic and inorganic. *Organic solids* in wastewater, simply put, are materials that are (or were) at one time alive and that will burn or volatilize at 550°C after 15 minutes in a muffle furnace (i.e., waste that is not exposed directly to flame, gases, or ash). The percent organic material within sludge will determine how unstable it is.

The inorganic material within sludge will determine how stable it is. The *inorganic solids* are those solids that were never alive and will not burn or volatilize at 550 °C after 15 minutes in a muffle furnace. Inorganic solids are generally not subject to breakdown by biological action and are considered stable. Certain inorganic solids, however, can create problems when related to the environment, for example, heavy metals such as copper, lead, zinc, mercury, and others. These can be extremely harmful if discharged.

Organic solids may be subject to biological decomposition in either an aerobic or anaerobic environment. Decomposition of organic matter (with its production of objectionable byproducts) and the possibility of toxic organic solids within the sludge compound the problems of sludge disposal.

The pathogens in domestic sewage are primarily associated with insoluble solids. Primary wastewater treatment processes concentrate these solids into sewage sludge, so untreated or raw primary sewage sludges have higher quantities of pathogens than the incoming wastewater. Biological wastewater treatment processes such as lagoons, trickling filters, and activated sludge treatment may substantially reduce the number of pathogens in wastewater (USEPA, 1989). These processes may also reduce the number of pathogens in sewage sludge by creating adverse conditions for pathogen survival.

Nevertheless, the resulting biological sewage sludges may still contain sufficient levels of pathogens to pose a public health and environmental concern. Moreover, insects, birds, rodents, and domestic animals may transport sewage sludge and pathogens from sewage sludge to humans and to animals. Vectors are attracted to sewage sludge as food source, and the reduction of the attraction of vectors to sewage sludge to prevent the spread of pathogens is a focus of current regulations. Sludge-borne pathogens and vector attraction are discussed in the following section.

SLUDGE PATHOGENS AND VECTOR ATTRACTION

A pathogen is an organism capable of causing disease. Pathogens infect humans through several different pathways including ingestion, inhalation, and dermal contact. The infective dose, or the number of pathogenic organisms to which a human must be exposed to become infected, varies depending on the organism and on the health status of the exposed individual.

Pathogens that propagate in the enteric or urinary system of humans and are discharged in feces or urine pose the greatest risk to public health with regard to the use and disposal of sewage sludge. Pathogens are also found in the urinary and enteric systems of other animals and may propagate in non-enteric settings.

As mentioned earlier, the four major types of human pathogenic (disease-causing) organisms [bacteria, viruses (including COVID-19), protozoa, and helminths (worms)] all may be present in domestic sewage. The actual species and quantity of pathogens present in the domestic sewage from a particular municipality (and the sewage sludge produced when treating the domestic wastewater) depend on the health status of the local community and may vary substantially at different times. The level of pathogens present in treated sewage sludge (biosolids) also depends on the reductions achieved by the wastewater and sewage sludge treatment processes.

If improperly treated sewage sludge were illegally applied to land, or placed on a surface disposal site, humans and animals could be exposed to pathogens directly by coming into contact with sewage sludge, or indirectly by consuming drinking water or food contaminated by sewage sludge pathogens, insects, birds, rodents, and even farm workers could contribute to these exposure routes by transporting sewage sludge and sewage sludge pathogens away from the site. Potential routes of exposure include:

Direct Contact
- Touching the sewage sludge.
- Walking through an area—such as a filed, forest, or reclamation area—shortly after sewage sludge application.

- Handling soil from fields where sewage sludge has been applied.
- Inhaling microbes that become airborne (via aerosols, dust, etc.) during sewage sludge spreading or by strong winds, plowing or cultivating the soils after application.

Indirect Contact

- Consumption of pathogen-contaminated crops grown on sewage sludge-amended soil or of other food products that have been contaminated by contact with these crops or field workers, etc.
- Consumption of pathogen-contaminated milk or other food products from animal contaminated by grazing in pastures or fed crops grown on sewage sludge-amended fields.
- Ingestion of drinking water or recreational waters contaminated by runoff from nearby land application sites or by organisms from sewage sludge migrating into groundwater aquifers.
- Consumption of inadequately cooked or uncooked pathogen-contaminated fish from water contaminated by runoff from a nearby sewage sludge application site.
- Contact with sewage sludge or pathogens transported away from the land application or surface disposal site by rodents, insects, or other vectors, including grazing animals or pets.

DID YOU KNOW?

The purpose of USEPA's Part 503 regulation is to place barriers in the pathway of exposure either by reducing the number of pathogens in the treated sewage sludge (biosolids) to below detectable limits, in the case of Class A treatment, or, in the case of Class B treatment, by preventing direct or indict contact with any pathogens possibly present in the biosolids. Each potential pathway has been studied to determine how the potential for public health risk can be alleviated.

One of the lesser impacts to public health can be from inhalation of airborne pathogens. Pathogens may become airborne via the spray of liquid biosolids from a splash plate or high-pressure hose, or in fine particulate dissemination as dewatered biosolids are applied or incorporated. While high-pressure spray applications may result in some aerosolization of pathogens, this type of equipment is generally used on large, remote sites such as forests, where the impact on the public is minimal. Fine particulates created by the application of dewatered biosolids or the incorporation of biosolids into soil may cause very localized fine particulate/dusty conditions, but particles in dewatered biosolids are too large to travel far, and the fine particulates do not spread beyond the immediate area. The activity of applying and incorporating biosolids may create dusty conditions. However, the biosolids are moist materials and do not add to the dusty condition, and by the time biosolids have dried sufficiently to create fine particulates, the pathogens have been reduced (Yeager and Ward, 1981).

With regard to vector attraction reduction, it can be accomplished in two ways: by treating the sewage sludge to the point at which vectors will no longer be attracted to the sewage sludge and by placing a barrier between the sewage sludge and vectors.

THE BOTTOM LINE

This section summarizes the findings and conclusions drawn from research, investigation, and modeling procedures conducted by HRSD, USGS, and CH2M for HRSD's SWIFT project are provided. These summarized procedures and findings are based on the evaluation of injection well rates, WWTP injection capacities, and the hydraulic response of the Potomac Aquifer System beneath the HRSD service area to injection operations. In addition, these bottom-line conclusions are from HRSD's (2016b) *Sustainable Water Recycling Initiative: Groundwater Injection Geochemical Compatibility Feasibility Evaluation. Report 1.* Virginia Beach, VA. Hampton Roads Sanitation District. Compiled by CH2M Newport News, VA and are summarized as follows:

- The transmissivities and available head for injection in the Potomac Aquifer System beneath each HRSD WWTP appear to support individual injection well capacities ranging between 3 and 8 MGD.
- In adhering to practical well design standards (such as, borehole and casing diameter and pumping capacities), injection capacities were capped at 3 MGD for this evaluation.
- To account for maintenance necessary for injection wells screened in sandy aquifers, one additional injection well was added in every five required to meet the effluent disposal rate at each WWTP.
- Accordingly, the number of injection wells ranged from 5 at Army Base and Williamsburg to 17 at the Virginia Initiative Plant.
- Analytical groundwater flow modeling indicated that of the seven WWTP sties tested,
- Army Base, James River, Williamsburg, and York River were able to meet the 2040 projected demands, within the site boundaries.
- Only Army Base and Williamsburg met the demands using the original two-aquifer approach.
- Conditions at James River and York River required screening all three of the Potomac Aquifer System aquifers to meet the 2014 demands.
- Sensitivity testing at York River Treatment Plant revealed that the modeled solution appeared sensitive to all parameters tested (transmissivity, storage coefficient, injection rates, simulation duration, and static water levels).
- The modeled solution exhibited the greatest sensitivity to changes in transmissivity. Changes in static water level resulted in increasing or decreasing the modeled head by the magnitude in the change of the water level.
- An evaluation of hydraulic interference between two wells at the York River WWTP reveal significant interference between wells spaced even 3,000 feet away.

- The results of the analytical modeling show that hydraulic interference exerts significant influence over the feasibility for replenishing the aquifer at the project 2040 rates.
- Injection was successfully simulated using the VCPM at each WWTP except York River. Because of its location inside the outer rim of the Chesapeake Bay Bolide Impact Crater, the VCPM simulates very low coefficients of hydraulic conductivity for the Potomac Aquifer System beneath the York River Treatment Plant.
- Injection at each WWTP resulted in removing most of the critical cells and region-wide, recovering water levels in the Potomac Aquifer System.
- Water levels in all injection scenarios resulted in simulated water levels that exceeded the land surface across the HRSD service area.

THE BOTTOM LINE ON HUMAN WASTE SOLIDS

The solid part of human waste, excreta, is a waste that has the potential of being morphed into a beneficial reuse product; thus, the title of "waste" can be eliminated if the sludge/biosolids are properly processed/treated and applied. Note that both sides, negative and positive aspects of biosolids usage have been presented in this section because it is important to point out that raw sewage solids have unhealthy attributes that must be guarded against, no matter their disposition. However, because of the negative aspects of raw, untreated, and not properly disposed of human waste solids, turning it into a beneficial reuse product makes good sense.

THE BOTTOM LINE

Modeling and geochemical evaluation resulting from mixing native groundwater and injectate and the reactions between injectate and aquifer minerals in the Potomac Aquifer System beneath HRSD's York River Treatment have generated several conclusions. These bottom-line conclusions are from *Sustainable Water Recycling Initiative: Groundwater Injection Geochemical Compatibility Feasibility Evaluation. Report 2 0* by Hampton Roads Sanitation District, Virginia Beach Virginia, and its primary consultant CH2M Newport News, VA and are summarized as follows:

- The chemistry of reverse osmosis, a potential injectate, differed significantly from the chemistry of native groundwater exhibited by the Upper Potomac Aquifer, Middle Potomac Aquifer, and Lower Potomac Aquifer.
- Reverse osmosis displayed a dilute ionic strength that differed by over one order of magnitude from the chemistry encountered in the Upper Potomac Aquifer and Middle Potomac Aquifer, while approaching two orders of magnitude when compared against the Lower Potomac.
- Reverse osmosis exhibits a calcium-bicarbonate water chemistry, while groundwater from the three Potomac Aquifer System aquifers uniformly exhibited sodium-chloride chemistry.

- The low ionic strength of reverse osmosis compared to groundwater from the Potomac Aquifer System represents a concern for injection operations, particularly for its potential to disperse clay minerals. Once dispersed, clay particles migrate through connected pores in the aquifer until accumulating and blocking narrowed pores, reducing aquifer permeability and ultimately injection well capacity.
- A USGS-sponsored Aquifer Storage and Recovery facility tested at Norfolk in the 1970s used an injectate similar in ionic strength and cation chemistry to reverse osmosis. The injection capacity of the Aquifer Storage and Recovery well declined by 50% after only four hours of operation, dropping 75% over several days. USGS was not able to restore the capacity of the Aquifer Storage and Recovery well, despite applying several, for the time, state of the art rehabilitation techniques.
- Cores collected at the city of Chesapeake's aquifer storage and recovery facility exhibited trace concentrations of smectitic clays dispersed throughout the interstices of every sample collected in aquifer sands. Smectites possess a complex lattice expanding structure vulnerable to dispersion or swelling when exposure to dilute water.
- In considering the varied cation chemistry between reverse osmosis and groundwater in the Potomac Aquifer System aquifers, the doubly charged calcium ion should benefit the long-term stability of clay minerals where calcium exchanges for sodium. However, calcium, when hydrated, exhibits a large ionic radius that can damage clay minerals upon entering the position vacated left by sodium, fragmenting the edges of the mineral, and mobilizing the fragments in the aquifer environment.
- Conversely, cores from the city of Chesapeake project, analyzed for cation exchange capacity displayed little tendency to exchange, which is a benefit of injection operations.
- Geochemical modeling of potential clay minerals in the Potomac Aquifer Systems aquifers produced a similar result with the stability of clay minerals improving over time during injection operations.
- Given the concerns with reverse osmosis as a source of injectate, and the problems experienced at the City of Norfolk's Aquifer Storage and Recovery project, HRSD should consider eliminating reverse osmosis from further evaluation on the SWIFT project.
- The ionic strength of nanofiltration and biologically activated carbon injectate fell within one order of magnitude of the groundwater chemistries originating from the Potomac Aquifer System aquifers. Nanofiltration and biological activated carbon as a source if injectate, represent significantly less of a concern of dispersing water-sensitive clays in the Potomac Aquifer System aquifers during injection operations.
- Applying a calcium or aluminum chloride treatment to the Potomac Aquifer System aquifers before initiating injection operations, offers a viable alternative for stabilizing clay minerals in situ, precluding formation damage, and injectivity losses should regulators selected reverse osmosis as the most

viable method for protecting local water users. These treatments could also benefit injection operations using nanofiltration of biologically activated carbon as the preferred injectate.

- Biologically activated carbon exhibits 0.7 mg/L iron, which presents a considerable source of TSS in the injectate, and a strong physical plugging agent in injection wells. HRSD will need to remove iron from biologically activated carbon effluent before employing it as an injectate.
- Geochemical modeling runs, simulating mixing between reverse osmosis, nanofiltration, and biologically activated carbon and groundwater from the Potomac Aquifer System aquifers showed no evidence of deleterious reactions that might clog the injection wells or surrounding aquifer such as precipitating oxide, hydroxide, carbonate, or sulfate minerals, or the dissolution of silicate minerals.
- Geochemical modeling, simulating the mixing between the three groundwaters in an injection well screening the Upper Potomac Aquifer, Middle Potomac Aquifer, and Lower Potomac Aquifer displayed no evidence of deleterious reactions that might clog the injection well or surrounding aquifers.
- Cores collected at the city of Chesapeake contained the iron carbonate mineral siderite at amounts ranging from 0.5% to 19% of the whole rock composition. In reactions with injectate-containing dissolved oxygen, siderite released up to 130 mg/L Fe II.
- Although not a concern for injection operations, dissolving siderite can compromise the quality of the disposed water, prompting attention from state and federal regulators.
- Adjusting the pH of the injectate water with a source of hydroxyl like sodium or potassium hydroxide can help lower Fe II concentrations. During model runs simulating reactions between injectate containing varying amounts of dissolved oxygen and pH of 8.5 with pyrite. Fe II concentrations fell below 10E-7 mg/L. Fe II oxidized to Fe III, which precipitated as Fe III-oxide and Fe III-hydroxide minerals.

REFERENCES AND RECOMMENDED READING

Anderman, E.R., and Hill, M.C. (2000). *MODFLOW 2000*. Water. USGS.gov/ogw/mudflow/MODFLOW.html.

Aulenbach, B.T. (2003). *Comparison of methods used to compare estuarian lake evaporation*. Lake Seminole, Atlanta, Georgia.

Bahremand, A., and De Smedt, F. (2008). Distributed hydrological modeling and sensitivity analysis in Torysa watershed, Slovakia. *Water Resources Management*, 22(3): 293–408.

Bair, E.S., Springer, A.E., and Roadcap, G.S. (1992). *CAPZONE*. Columbus, OH: Ohio State University.

Brown, D.L. and Silvey, W.D. (1977). *Same aquifer*. Norfolk, Virginia. Accessed 02/02/21 @ https://pubs.usgs.gov/pp/report/pdf.

Carlson, J. et al. (1992). Evidence of decreasing semen in past 50 years. *BMJ* 3052 609–612.

CH2M. (2016). *Sustainable Water Recycling Initiative: Groundwater Injection Geochemical Compatibility Feasibility Evaluation*, Report No. 1. Newport News, VA: CH2M.

Colburn, t., et al. (1993). Developmental effects of endocrine-disrupting chemicals in wildlife and humans. *Environ Health Perspect.* 101(5) 378–384.

Conlon, W.L. (1989). Membrane technology in Florida. *Journal of AWWA* 81(11) 43–46.

Crites, R.W. (2000). *Land treatment system.* NY: McGraw-Hill.

Davies, J.H. (1962). Surface complex modeling. *Minerology* 23 177–260.

Debye, P. and Huykel (1923). Phase equilibrium model. Accessed 02/14/21 @ https://chim .libr.texts.org.

Drever, J.I. (1988). *The geochemistry of natural water.* Upper saddle river, NJ: Prentice-Hall.

Dzomback, D.A. and Morel, F.M.M. (1990). *Surface complexion modeling.* NY: John Wiley.

Eggleston, J., and Pope, J. (2013). Land subsidence and relative sea-level rise in the southern Chesapeake Bay region: U.S. *Geological Survey Circular,* 1392, 30. doi:10.3111/ cir1392.

Evangelou, V.P. (1995). *Pyritex Oxidation and its control.* NY: CRC Press.

Eykoff, P. (1974). *System Identification Parameter and State Estimation.* New York: Wiley & Sons.

Goigel, J. (1991). Hydraulic wastewater loading rates. Accessed 02/04/21 @ https://soilks .wise/aides.

Gordon, J. (1984). Incineration and treatment of hazardous waste. Accessed 02/12/21@ https://nepis.epa.gov/exe/zypurl.Cgl?dockey.

Gray, D.H. and Rex, R.W. (1966). Gray and Rex electro kinetic process. Accessed 02/12/21 @ Buy.com/video.

Hamilton, P.A., and Larson, J.D. (1988). *Hydrogeology and Analysis of the Ground-Water-Flow System in the Coastal Plain of Southeastern Virginia.* U.S. Geological Survey Water Resources Investigations Report 87-4240, Richmond, VA.

Hantash, J.E., and Jacob, C.E. (1955). Nonsteady radial flow in an infinite leaky aquifer. *Transactions of the American Geophysical Union,* 36(1): 95–100.

Hewitt, C.H. (1963). Analytical techniques for recognizing water sensitive reservoir rocks. *Jour. Petrol. Tech.* No 16 793–818.

Heywood, C.E., and Pope, J.P. (2009). Simulation of groundwater flow in the Coastal plain aquifer system of Virginia, Scientific Investigation Report 2009–5039. U.S. Geological Survey. Water.usgs.gov/ogw/seawat.

Hill, M., Kavetski, D., Clark, M., Ye, M., Arabic, M., Lu, D., Foglia, L., and Mehl, S. (2015). Practical use of computationally frugal model analysis methods. *Groundwater,* 54(2): 159–170.

Hill, M., and Tiedeman, C. (2007). *Effective Groundwater Model Calibration, with Analysis of Data, Sensitivities, Prediction, and Uncertainty.* New York: John Wiley & Sons.

Holdahl, S.R., and Morrison, N. (1974). Regional investigations of vertical crustal movements in the U.S., using precise relevelings and mareograph data. *Tectonophysics,* 23(4): 373–390.

HRSD. (2016). *SWIFT (Sustainable Water Initiative for Tomorrow).* Accessed 02/12/2021 @ www.hrsd.com.

HRSD. (2016). *Sustainable Water Recycling Initiative: Groundwater Injection Hydraulic Feasibility Evaluation.* Virginia Beach, VA: Hampton Roads Sanitation District. Compiled by CH2M Newport News, VA.

Jones, B.D. (1980). *Service Delivery in the City: Citizen's Demand and Bureaucratic Rules.* New York: Longman, p. 2.

Kavlock et al., (1996). Research needs for the risk assessment of health and environmental effects of endocrine disruptors. *Environ health Perspect.* 104(suppl4) 715–740.

Laczniak, R.J., and Meng, III, A.A. (1988). *Ground-Water Resources of the York-James Peninsula of Virginia.* U.S. Geological Survey Water Resources Investigations Report 88-4059, Richmond, VA.

Langevin, C.D., Thorne, D.T., Jr., Dausman, A.M., Sukop, M.S., and Guo, Weixing. (2008). SEWAT Version. U.S. Geological Survey. Accessed 02/12/2021 @ https://pubs.usgs.gov/tm/tmbaza//.

Langmuir, D. (1997). *Aqueous environmental geochemistry.* Upper Saddle River, NJ: Prentice-Hall.

Leamer, E. (1978). *Specification Searches: Ad Hoc Inferences with Nonexperimental Data.* New York: John Wiley & Sons.

Lee, T.M. and Swancar, A. (1996). *Influence of evaporation Lake Lucerne.* Polk county, FL.: United States Geological Survey.

Mainer, O.E. (1923). *Introduction to geology.* New York: Harper-Collins.

Meade, R.H. (1964). Effect of compaction of some properties of sediments. *Developments in Sedimentology* 16 123–129.

McFarland, E.R. (2013). *Sediment Distribution and Hydrologic Conditions of the Potomac Aquifer in Virginia and Parts of Maryland and North Caroline.* United States Geological Survey Scientific Investigations report 2013–5116, Reston, VA.

McFarland, E.R., and Bruce, T.S. (2006). *The Virginian Coastal Plain Hydrogeologic Framework.* United States Geologic Survey Professional paper 1731, Reston, VA.

McGill, K., and Lucas, M.C. (2009). *Mitigating Specific Capacity Losses in Aquifer Storage and Recovery Wells in the New Jersey Coastal Plain.* New Jersey Chapter of American Water Works Association, Annual Conference, Atlantic City, NJ.

Meng, A.A., and Harsh, J.F. (1988). *Hydrogeologic Framework of the Virginia Coastal Plain.* United States geological Survey Professional Paper 1404-C, Washington, DC.

Miller, G.T. (1989). *Environmental science: An introduction.* Belmont, CA: Wadsworth Publishing Company.

Office of Saline Water (1970). *Brine disposal manual.* Washington, DC: US Dept of Interior.

Parkhurst, D.L. (1995). User's guide to PHREEQC computer program. Accessed 02/02/21 @ https://usgs.gov/wri/1995.

Pannell, D.J. (1997). Sensitivity analysis of normative economic models: Theoretical framework and practical strategies. *Agricultural Economics,* 16: 139–152.

Pope, J.P, and Burbey, T.J. (2004). Multiple-aquifer characteristics from single borehole extensometer records. *Ground Water,* 42(1): 45–58.

Pyne, D.G. (1995). *Groundwater Recharge and Wells.* Ann Arbor, MI: Lewis Publishers.

Pyne, D.G. (2005). *Aquifer Storage and Recovery: A Guide to Groundwater Recharge through Wells.* Gainesville, FL: ASR Press.

Reclamation (1969). *Evaporation from large water bodies.* Washington DC: Bureau of Reclamation.

Reed, M.G. (1972). Clay dispersion. Effect of compaction of some properties of sediments. *Developments in Sedimentology* 16 131–139.

Rosenberry, D.O. et al. (1993). Multi-lake evaporation: Energy budget. Accessed 03/01/21 @ https://www.sciencedirect/science/article.

Smith, B.S. (1999). *The Potential for Saltwater Intrusion in the Potomac Aquifers of the York-James Peninsula.* U.S. Geological survey Water Resources Investigations Report 98-4187, Richmond, VA.

Strycker, L. and Collins, A. (1987). Injection of Haz waste into deep wells. Accessed 02/02/2021 @https://nepis.epa.gov/exe.

Stuyzfand, P.J. (1993). *Hydro chemistry and hydrology of the coastal dune area of Netherlands.* Amsterdam, Netherlands: Free U.

Theis, C.V. (1935). The relation between lowering of the piezometric surface and the rate of duration of discharge of a well groundwater storage. *Transactions of the America Geophysical Union,* 16(2): 111–114.

Treifke, R.H. (1973). *Geologic Studies Coastal Plain of Virginia*, Bulletin 83. Richmond, VA: Virginia Division of Mineral Resources.

Truesdell, A.H. and Jones, B.F. (1974). *WATEQ, Computer program*. Washington, DC: US Geological Survey.

USEPA (2006). *Biphenyls in ecological risk assessment*. Epa/1001r-05;004. Washington, DC: United States Environmental Protection Agency.

USEPA. (2016). *Aquifer Recharge and Aquifer Storage and Recovery*. Accessed 02/12/2021 @ https://epa.gov/uic/aquifer-recharge-and-aquifer-storage-and-recovery.

Virginia Department of Environmental Quality (VADEQ). (2006a). *Status of Virginia's Water Resources*. Richmond, VA.

Virginia Department of Environmental Quality (VADEQ). (2006b). *Virginia Coastal Palin Model 2005 Withdrawals Simulation*. Richmond, VA.

Warner, D.L., and Lehr, J. (1981). *Subsurface Wastewater Injection, The Technology of Injecting Wastewater into Deep Wells for Disposal*. Berkeley, CA: Premier Press.

Yeager, J.G. and Ward, R.L. (1981). Effects of moisture on bacteria in wastewater sludge. *Applied Env. Micro.* 41(5) 1117–1122.

Glossary

A

Absorption: any process by which one substance penetrates the interior of another substance.

Acid: has a pH of water less than 5.5; pH modifier used in the U.S. Fish and Wildlife Service wetland classification system; in common usage, acidic water has a pH less than 7.

Acidic deposition: the transfer of acidic or acidifying substances forms the atmosphere to the surface of the Earth or to objects on its surface. Transfer can be either by wet-deposition processes (rain, snow, dew, fog, frost, hail) or by dry deposition (gases, aerosols, or fine-to-coarse particles).

Acid rain: precipitation with higher-than-normal acidity, caused primarily by sulfur and nitrogen dioxide air pollution.

Acre-foot (acre-ft.): the volume of water needed to cover an acre of land to a depth of one foot; equivalent to 43,560 cubic feet or 32,851 gallons.

Activated carbon: a very porous material that after being subjected to intense heat to drive off impurities can be used to adsorb pollutants from water.

Adsorption: the process by which one substance is attracted to and adheres to the surface of another substance, without actually penetrating its internal structure.

Aeration: a physical treatment method that promotes biological degradation of organic matter. The process may be passive (when waste is exposed to air), or active (when a mixing or bubbling device introduces the air).

Aerobic bacteria: a type of bacteria that requires free oxygen to carry out the metabolic function.

Algae: chlorophyll-bearing nonvascular, primarily aquatic species that have no true roots, stems, or leaves; most algae are microscopic, but some species can be as large as vascular plants.

Algal bloom: the rapid proliferation of passively floating, simple plant life, such as blue-green algae, in and on a body of water.

Alkaline: has a pH greater than 7; pH modifier in the US Fish and Wildlife Service wetland classification system; in common usage, a pH of water greater than 7.4.

Alluvial aquifer: a water-bearing deposit of unconsolidated material (sand and gravel) left behind by a river or other flowing water.

Alluvium: general term for sediments of gravel, sand, silt clay, or other particulate rock material deposited by flowing water, usually in the beds of rivers and streams, on a flood plain, on a delta, or at the base of a mountain.

Alpine snow glade: a marshy clearing between slopes above the timberline in mountains.

Amalgamation: the dissolving or blending of a metal (commonly gold and silver) in mercury to separate it from its parent material.

Ammonia: a compound of nitrogen and hydrogen (NH_3) that is a common byproduct of animal waste. Ammonia readily converts to nitrate in soils and streams.

Anaerobic: pertaining to, taking place in, or caused by the absence of oxygen.

Anomalies: as related to fish, externally visible skin or subcutaneous disorders, including deformities, eroded fins, lesions, and tumors.

Anthropogenic: having to do with or caused by humans.

Anticline: a fold in the Earth's crust, convex upward, whose core contains stratigraphically older rocks.

Aquaculture: the science of farming organisms that live in water, such as fish, shellfish, and algae.

Aquatic: living or growing in or on water.

Aquatic guidelines: specific levels of water quality which, if reached, may adversely affect aquatic life. These are non-enforceable guidelines issued by a governmental agency or other institution.

Aquifer: a geologic formation, group of formations, or part of a formation that contains sufficient saturated permeable material to yield significant quantities of water to springs and wells.

Aquitard: A saturated, but poorly permeable, geologic unit that impedes groundwater movement and does not yield water freely to wells, but which may transmit appreciable water to and from adjacent aquifers and, where sufficiently thick, may constitute an important groundwater storage unit. Really extensive aquitards may function regionally as confined units within aquifer systems.

Arroyo: a small, deep, flat-floored channel or gully of an ephemeral or intermittent stream, usually with nearly vertical banks cut, into unconsolidated material.

Artesian: an adjective referring to confined aquifers. Sometimes the term artesian is used to denote a portion of a confined aquifer where the altitude of the potentiometric surface is above the land surface (flowing wells and artesian wells are synonymous in this usage). But more generally the term indicates that the altitudes of the potentiometric surface are above the altitude of the base or the confining unit (artesian wells and flowing wells are not synonymous in this case.

Artificial recharge: augmentation of natural replenishment of groundwater storage by some method of construction, spreading of water, or by pumping water directly into an aquifer.

Atmospheric deposition: the transfer of substances from the air to the surface of the Earth, either in wet form (rain, fog, snow, dew, frost, hail) or in dry form (gases, aerosols, particles).

Atmospheric pressure: the pressure exerted by the atmosphere on any surface beneath or within it; equal to 14.7 pounds per square inch at sea level.

Average discharge: as used by the US Geological Survey, the arithmetic average of all complete water years of record of surface water discharge whether consecutive or not. The term "average" generally is reserved for an average of record and "mean" is used for averages of shorter periods, namely, daily, monthly, or annual mean discharges.

B

Background concentration: a concentration of a substance in a particular environment that is indicative of minimal influence by human (anthropogenic) sources.

Backwater: a body of water in which the flow is slowed or turned back by an obstruction such as a bridge or dam, an opposing current, or the movement of the tide.

Bacteria: single-celled microscopic organisms.

Bank: the sloping ground that borders a stream and confines the water in the natural channel when the water level, or flow, is normal.

Bank storage: the change in the amount of water stored in an aquifer adjacent to a surface-water body resulting from a change in the stage of the surface-water body.

Barrier bar: an elongate offshore ridge, submerged at least at high tide, built up by the action of waves or currents.

Base flow: the sustained low flow of a stream, usually groundwater inflow to the stream channel.

Basic: the opposite of acidic; water that has a pH of greater than 7.

Basin and range physiography: a region characterized by a series of generally north-trending mountain ranges separated by alluvial valleys.

Bed material: sediment comprising the streambed.

Bed sediment: the material that temporarily is stationary in the bottom of a stream or other watercourse.

Bedload: sediment that moves on or near the streambed and is in almost continuous contact with the bed.

Bedrock: a general term used for solid rock that underlies soils or other unconsolidated material.

Benthic invertebrates: insects, mollusk, crustaceans, worms, and other organisms without a backbone that live in, on, or near the bottom of lakes, streams, or oceans.

Benthic organism: a form of aquatic life that lives on or near the bottom of streams, lakes, or oceans.

Bioaccumulation: the biological sequestering of a substance at higher concentrations than that at which it occurs in the surrounding environment or medium. Also, the process whereby a substance enters organisms through the gills, epithelial tissues, dietary, or other sources.

Bioavailability: the capacity of a chemical constituent to be taken up by living organisms either through physical contact or by ingestion.

Biochemical: refers to chemical processes that occur inside or are mediated by living organisms.

Biochemical process: a process characterized by, produced by, or involving chemical reactions in living organisms.

Biochemical oxygen demand (BOD): the amount of oxygen required by bacteria to stabilize decomposable organic matter under aerobic conditions.

Biodegradation: the transformation of a substance into new compounds through biochemical reactions or the actions of microorganisms such as bacteria.

Biological treatment: a process that uses living organisms to bring about chemical changes.

Biomass: the amount of living matter, in the form of organisms, present in a particular habitat, usually expressed as weight-per-unit area.

Biota: all living organisms of an area.

Blue hole: a subsurface void, usually a solution sinkhole, developed in carbonate rocks that are open to the Earth's surface and contains tidally influenced waters of fresh, marine, or mixed chemistry.

Bog: a nutrient-poor, acidic wetland dominated by a waterlogged, spongy mat of sphagnum moss that ultimately forms a thick layer of acidic peat; generally has no inflow or outflow; fed primarily by rain water.

Brackish water: water with a salinity intermediate between seawater and freshwater (containing from 1,000 to 10,000 milligrams per liter of dissolved solids).

Breakdown product: a compound derived by chemical, biological, or physical action upon a pesticide. The breakdown is a natural process that may result in a more toxic or a less toxic compound and a more persistent or less persistent compound.

Breakpoint chlorination: the addition of chlorine to water until the chlorine demand has been satisfied and free chlorine residual is available for disinfection.

C

Calcareous: a rock or substance formed of calcium carbonate or magnesium carbonate by biological deposition or inorganic precipitation or containing those minerals in sufficient quantities to effervesce when treated with cold hydrochloric acid.

Capillary fringe: the zone above the water table in which water is held by surface tension. Water in the capillary fringe is under a pressure less than atmospheric.

Carbonate rocks: rocks (such as limestone or dolostone) that are composed primarily of minerals (such as calcite and dolomite) containing a carbonate ion.

Cenote: steep-walled natural well that extends below the water table; generally caused by the collapse of a cave roof; term reserved for features found in the Yucatan Peninsula of Mexico.

Center pivot irrigation: an automated sprinkler system involving a rotating pipe or boom that supplies water to a circular area of an agricultural field through sprinkler heads or nozzles.

Channel scour: erosion by flowing water and sediment on a stream channel; results in removal of mud, silt, and sand on the outside curve of a stream bend and the bed material of a stream channel.

Channelization: the straightening and deepening of a stream channel to permit the water to move faster or to drain a wet area for farming.

Chemical treatment: a process that results in the formation of a new substance or substances. The most common chemical water treatment processes include coagulation, disinfection, water softening, and filtration.

Chlordane: octachlor-4,7-methanotetrahydroindane. An organochlorine insecticide no longer registered for use in the US. Technical chlordane is a mixture in which the primary components are cis- and trans-chlordane, cis- and trans-nonachlor, and heptachlor.

Chlorinated solvent: a volatile organic compound containing chlorine. Some common solvents are trichloroethylene, tetrachloroethylene, and carbon tetrachloride.

Chlorofluorocarbons: a class of volatile compounds consisting of carbon, chlorine, and fluorine. Commonly called freons, which have been in refrigeration mechanisms, as blowing agents in the fabrication of flexible and rigid foams, and, until banned from use several years ago, as propellants in spray cans.

Chlorination: the process of adding chlorine to the water to kill disease-causing organisms or to act as an oxidizing agent.

Chlorine demand: a measure of the amount of chlorine that will combine with impurities and is therefore unavailable to act as a disinfectant.

Cienaga: a marshy area where the ground is wet due to the presence of seepage of springs.

Clean Water Act (CWA): federal law dating to 1972 (with several amendments) with the objective to restore and maintain the chemical, physical, and biological integrity of the nation's waters. Its long-range goal is to eliminate the discharge of pollutants into navigable waters and to make national waters fishable and swimmable.

Climate: the sum total of the meteorological elements that characterize the average and extreme conditions of the atmosphere over a long period of time at any one place or region of the Earth's surface.

Coagulants: chemicals that cause small particles to stick together to form larger particles.

Coagulation: a chemical water treatment method that causes very small, suspended particles to attract one another and form larger particles. This is accomplished by the addition of a coagulant that neutralizes the electrostatic charges that cause particles to repel each other.

Coliform bacteria: a group of bacteria predominantly inhabiting the intestines of humans or animals, but also occasionally found elsewhere. The presence of the bacteria in the water is used as an indication of fecal contamination (contamination by animal or human wastes).

Color: a physical characteristic of water. Color is most commonly tan or brown from oxidized iron, but contaminants may cause other colors, such as green or blue. Color differs from turbidity, which is water's cloudiness.

Combined sewer overflow: a discharge of untreated sewage and stormwater to a stream when the capacity of a combined storm/sanitary sewer system is exceeded by storm runoff.

Communicable diseases: Usually caused by *microbes*—microscopic organisms including bacteria, protozoa, and viruses. Most microbes are essential components of our environment and do not cause disease. Those that do are called pathogenic organisms, or simply *pathogens*.

Community: in ecology, the species that interact in a common area.

Community water system: a public water system that serves at least 15 service connections used by year-round residents, or regularly serves at least 25 year-round residents.

Compaction: in this book, compaction is used in its geologic sense and refers to the inelastic compression of the aquifer system. Compaction of the aquifer system reflects the rearrangement of the mineral grain pore structure and largely nonrecoverable reduction of the porosity under stress greater than the preconsolidation stress. Compaction, as used here, is synonymous with the term "virgin consolidation" used by soil engineers. The term refers to both the process and the measured change in thickness. As a practical matter, a very small amount (1–5%) of the compaction is recoverable as a slight elastic rebound of the compacted material if stresses are reduced.

Compaction, residual: is the compaction that would ultimately occur if a given increase in applied stress were maintained until steady-state pore pressures were achieved. Residual compaction may also be defined as the difference between (1) the amount of compaction that will occur ultimately for a given increase in applied stress, and (2) that which has occurred at a specified time.

Composite sample: a series of individual or grab samples taken at different times from the same sampling point and mixed together.

Compression: in this book, compression refers to the decrease in thickness of sediments, as a result of the increase in vertical compressive stress. Compression may be elastic (fully recoverable) or inelastic (nonrecoverable).

Concentration: the ratio of the quantity of any substance present in a sample of a given volume or a given weight compared to the volume or weight of the sample.

Cone of depression: the depression of heads around a pumping well caused by the withdrawal of water.

Confined aquifer (artesian aquifer): an aquifer that is completely filled with water under pressure and that is overlain by material that restricts the movement of water.

Confining bed: a layer of rock having very low hydraulic conductivity that hampers the movement of water into and out of an aquifer.

Confining layer: a body of impermeable or distinctly less permeable material stratigraphically adjacent to one or more aquifers that restricts the movement of water into and out of the aquifers.

Confining unit: a saturated, relatively low-permeability geologic unit is really extensive and serves to confine an adjacent artesian aquifer or aquifers. Leaky confining units may transmit appreciable water to and from adjacent aquifers.

Confluence: the flowing together of two or more streams; the place where a tributary joins the main stream.

Conglomerate: a coarse-grained sedimentary rock composed of fragments larger than 2 mm in diameter.

Consolidation: in soil mechanics, consolidation is the adjustment of a saturated soil in response to the increased load, involving the squeezing of water from the pores and decrease in void ratio or porosity of the soul. In this book, the geologic term "compaction" is used in preference to consolidation.

Constituent: a chemical or biological substance in water, sediment, or biota that can be measured by an analytical method.

Consumptive use: the quantity of water that is not available for immediate rescue because it has been evaporated, transpired, or incorporated into products, plant tissue, or animal tissue.

Contact recreation: recreational activities, such as swimming and kayaking, in which contact with water is prolonged or intimate, and in which there is a likelihood of ingesting water.

Contaminant: a toxic material found as an unwanted residue in or on a substance.

Contamination: degradation of water quality compared to original or natural conditions due to human activity.

Contributing area: the area in a drainage basin that contributes water to streamflow or recharge to an aquifer.

Core sample: a sample of rock, soil, or other material obtained by driving a hollow tube into the undisturbed medium and withdrawing it with its contained sample.

Criterion: a standard rule or test on which a judgment or decision can be based.

Cross connection: any connection between safe drinking water and a nonpotable water or fluid.

CxT value: the product of the residual disinfectant concentration **C**, in milligrams per liter, and the corresponding disinfectant contact time **T**, in minutes. Minimum **CxT** values are specified by the Surface Water Treatment Rule, as a means of ensuring adequate kill or inactivation of pathogenic microorganisms in water.

D

Datum plane: a horizontal plane to which ground elevations or water surface elevations are referenced.

Deepwater habitat: permanently flooded lands lying below the deep-water boundary of wetlands.

Degradation products: compounds resulting from the transformation of an organic substance through chemical, photochemical, and/or biochemical reactions.

Denitrification: a process by which oxidized forms of nitrogen such as nitrate are reduced to form nitrites, nitrogen oxides, ammonia, or free nitrogen commonly brought about by the action of denitrifying bacteria and usually resulting in the escape of nitrogen to the air.

Detection limit: the concentration of a constituent or analyte below which a particular analytical method cannot determine, with a high degree of certainty, the concentration.

Diatoms: single-celled, colonial, or filamentous algae with siliceous cell walls constructed of two overlapping parts.

Direct runoff: the runoff entering stream channels promptly after rainfall or snowmelt.

Discharge: the volume of fluid passing a point per unit of time, commonly expressed in cubic feet per second, million gallons per day, gallons per minute, or seconds per minute per day.

Discharge area (groundwater): the area where subsurface water is discharged to the land surface, to surface water, or to the atmosphere.

Disinfectants-disinfection byproducts (D-DBPs): a term used in connection with state and federal regulations designed to protect public health by limiting the concentration of either disinfectants or the byproducts formed by the reaction of disinfectants with other substances in the water (such as trihalomethanes—THMs).

Disinfection: a chemical treatment method. The addition of a substance (e.g., chlorine, ozone, or hydrogen peroxide), which destroys or inactivates harmful microorganisms, or inhibits their activity.

Dispersion: the extent to which a liquid substance introduced into a groundwater system spreads as it moves through the system.

Dissociate: the process of ion separation that occurs when an ionic solid is dissolved in water.

Dissolved constituent: operationally defined as a constituent that passes through a 0.45-micrometer filter.

Dissolved oxygen (DO): the oxygen dissolved in water usually expressed in milligrams per liter, parts per million, or percent of saturation.

Dissolved solids: any material that can dissolve in water and be recovered by evaporating the water after filtering the suspended material.

Diversion: a turning aside or alteration of the natural course of a flow of water, normally considered physically to leave the natural channel. In some states, this can be a consumptive use direct from another stream, such as by livestock watering. In other States, a diversion must consist of such actions as taking water through a canal, pipe, or conduit.

Dolomite: a sedimentary rock consisting chiefly of magnesium carbonate.

Domestic withdrawals: water used for normal household purposes, such as drinking, food preparation, bathing, washing clothes and dishes, flushing toilets, and watering lawns and gardens. The water may be obtained from a public supplier or may be self-supplied. Also called residential water use.

Drainage area: the drainage area of a stream at a specified location is that area, measured in a horizontal plane, which is enclosed by a drainage divide.

Drainage basin: the land area drained by a river or a stream.

Drainage divide: the boundary between adjoining drainage basins.

Drawdown: the difference between the water level in a well before pumping and the water level in the well during pumping. Also, for flowing wells, the reduction of the pressure head as a result of the discharge of water.

Drinking water standards: water quality standards measured in terms of suspended solids, unpleasant taste, and microbes harmful to human health. Drinking water standards are included in state water quality rules.

Drinking water supply: any raw or finished water source that is or may be used as a public water system or as drinking water by one or more individuals.

Drip irrigation: an irrigation system in which water is applied directly to the root zone of plants by means of applicators (orifices, emitters, porous tubing, or perforated pipe) operated under low pressure. The applicators can be placed on or below the surface of the ground or can be suspended from supports.

Drought: a prolonged period of less-than-normal precipitation such that the lack of water causes a serious hydrologic imbalance.

E

Ecoregion: an area of similar climate, landform, soil, potential natural vegetation, hydrology, or other ecologically relevant variables.

Ecosystem: a community of organisms considered together with the nonliving factors of its environment.

Effluent: outflow from a particular source, such as a stream that flows from a lake or liquid waste that flows from a factory or sewage-treatment plant.

Effluent limitations: standards developed by the EPA to define the levels of pollutants that could be discharged into surface waters.

Electrodialysis: the process of separating substances in a solution by dialysis, using an electric field as the driving force.

Electronegativity: the tendency for atoms that do not have a complete octet of electrons in their outer shell to become negatively charged.

Ellipsoid, Earth: a mathematically determined three-dimensional surface obtained by rotating an ellipse about its semi-minor axis. In the case of the Earth, the ellipsoid is the modeled shape of its surface, which is relatively flattened in the polar axis.

Ellipsoid, height: the distance of a point above the ellipsoid measured perpendicular to the surface of the ellipsoid.

Emergent plants: erect, rooted, herbaceous plants that may be temporarily or permanently flooded at the base but do not tolerate prolonged inundation of the entire plant.

Enhanced Surface Water Treatment Rule (ESWTR): a revision of the original Surface Water Treatment Rule that includes new technology and requirements to deal with the newly identified problems.

Environment: the sum of all conditions and influences affecting the life of organisms.

Environmental sample: a water sample collected from an aquifer or stream for the purpose of chemical, physical, or biological characterization of the sampled resource.

Environmental setting: land area characterized by a unique combination of natural and human-related factors, such as row-crop cultivation or glacial-till soils.

Ephemeral stream: a stream or part of a stream that flows only in direct response to precipitation; it receives little or no water from springs, melting snow, or other sources; its channel is at all times above the water table.

EPT richness index: an index based on the sum of the number of taxa in three insect orders, Ephemeroptera (mayflies), Plecoptera (stoneflies), and Trichoptera (caddisflies), that are composed primarily of species considered to be relatively intolerant to environmental alterations.

Equipotential line: a line on a map or cross section along which total heads are the same.

Erosion: the process whereby materials of the Earth's crust are loosened, dissolved, or worn away and simultaneously moved from one place to another.

Eutrophication: the process by which water becomes enriched with plant nutrients, most commonly phosphorus and nitrogen.

Evaporite minerals (deposits): minerals or deposits of minerals formed by the evaporation of water-containing salts. These deposits are common in arid climates.

Evaporites: a class of sedimentary rocks composed primarily of minerals precipitated from a saline solution as a result of extensive or total evaporation of water.

Evapotranspiration: the process by which water is discharged to the atmosphere as a result of evaporation from the soil and surface-water bodies and transpiration by plants.

Exfoliation: the process by which concentric scales, plates, or shells of rock, from less than a centimeter to several meters in thickness, are stripped from the bare surface of a large rock mass.

F

Facultative bacteria: a type of anaerobic bacteria that can metabolize its food either aerobically or anaerobically.

Fall line: imaginary line marking the boundary between the ancient, resistant crystalline rocks of the Piedmont province of the Appalachian Mountains, and the younger, softer sediments of the Atlantic Coastal Plain province in the Eastern United States. Along rivers, this line commonly is reflected by waterfalls.

Fecal bacteria: microscopic single-celled organisms (primarily fecal coliforms and fecal streptococci) found in the wastes of warm-blooded animals. Their presence in water is used to assess the sanitary quality of water for body-contact recreation or for consumption. Their presence indicates contamination by the wastes of warm-blooded animals and the possible presence of pathogenic (disease-producing) organisms.

Federal Water Pollution Control Act (1972): The Act outlines the objective "to restore and maintain the chemical, physical, and biological integrity of the nation's waters." This 1972 act and the subsequent Clean Water Act amendments are the most far-reaching water pollution control legislation ever enacted. They provided comprehensive programs for water pollution control, uniform laws, and interstate cooperation. They provided grants for research, investigations, training, and information on national programs on surveillance, the effects of pollutants, pollution control, and the identification and measurement of pollutants. Additionally, they allot grants and loans for the construction of treatment works. The Act established national discharge standards with enforcement provisions.

The Federal Water Pollution Control Act established several milestone achievement dates. It required secondary treatment of domestic waste by publicly owned treatment works (POTWs), and application of "best practicable" water pollution control technology by industry by 1977. Virtually, all industrial sources have achieved compliance (because of economic difficulties and cumbersome federal requirements, certain POTWs obtained an extension to 1 July 1988 for compliance). The Act also called for new levels of technology to be imposed during the 1980s and 1990s, particularly for controlling toxic pollutants.

The Act mandates a strong pretreatment program to control toxic pollutants discharged by industry into POTWs. The 1987 amendments require that stormwater from the industrial activity must be regulated.

Fertilizer: any of a large number of natural or synthetic materials, including manure and nitrogen, phosphorus, and potassium compounds, spread on or worked into the soil to increase its fertility.

Filtrate: liquid that has been passed through a filter.

Filtration: a physical treatment method for removing solid (particulate) matter from water by passing the water through porous media such as sand or a man-made filter.

Flocculation: the water treatment process following coagulation; it uses gentle stirring to bring suspended particles together so that they will form larger, more settleable clumps called floc.

Flood: any relatively high streamflow that overflows the natural or artificial banks of a stream.

Flood attenuation: a weakening or reduction in the force or intensity of a flood.

Flood irrigation: the application of irrigation water whereby the entire surface of the soil is covered by ponded water.

Flood plain: a strip of relatively flat land bordering a stream channel that is inundated at times of high water.

Flow line: the idealized path followed by particles of water.

Flow net: The grid pattern formed by a network of flow lines and equipotential lines.

Flowpath: an underground route for ground-water movement, extending from a recharge (intake) zone to a discharge (output) zone such as a shallow stream.

Fluvial: pertaining to a river or stream.

Freshwater: water that contains less than 1,000 milligrams per liter of dissolved solids.

Freshwater chronic criteria: the highest concentration of a contaminant that freshwater aquatic organisms can be exposed to for an extended period of time (4 days) without adverse effects.

G

Geodetic datum: a set of constants specifying the coordinate system used for geodetic control, for example, for calculating the coordinates of points on the Earth.

Geoid, Earth: The sea-level equipotential surface or figure of the Earth. If the earth were completely covered by a shallow sea, the surface of this sea would conform to the geoid shaped by the hydrodynamic equilibrium of the water subject to gravitational and rotational forces. Mountains and valleys are departures from the reference geoid.

Grab sample: a single water sample collected at one time from a single point.

Groundwater: the fresh water found under the Earth's surface, usually in aquifers. Groundwater is a major source of drinking water, and a source of a growing concern in areas where leaching agricultural or industrial pollutants, or substances from leaking underground storage tanks are contaminating groundwater.

H

Habitat: the part of the physical environment in which a plant or an animal lives.

Hardness: a characteristic of water caused primarily by the salts of calcium and magnesium. It causes deposition of scale in boilers, damage in some industrial processes, and sometimes objectionable taste. It may also decrease soap's effectiveness.

Head: the height above a datum plane of a column of water. In a groundwater system, it is composed of the elevation head and pressure head.

Headwaters: the source and upper part of a stream.

Hydraulic conductivity: the capacity of a rock to transmit water. It is expressed as the volume of water at the existing kinematic viscosity that will move in unit time under a unit hydraulic gradient through a unit area measured at right angles to the direction of flow.

Hydraulic gradient: the change of hydraulic head per unit of distance in a given direction.

Hydrocompaction: the process of volume decrease and density increase that occurs when certain moisture-deficient deposits compact as they are wetted for the first time since burial. The vertical downward movement of the land surface that results from this process has also been termed "shallow subsidence" and "near-surface subsidence."

Hydrogen bonding: the term used to describe the weak but effective attraction that occurs between polar covalent molecules.

Hydrograph: graph showing the variation of water elevation, velocity, streamflow, or other property of water with respect to time.

Hydrologic cycle: literally, the water–earth cycle. The movement of water in all three physical forms through the various environmental mediums (air, water, biota, and soil).

Hydrology: the science that deals with water as it occurs in the atmosphere, on the surface of the ground, and underground.

Hydrostatic pressure: the pressure exerted by the water at any given point in a body of water at rest.

Hygroscopic: a substance that readily absorbs moisture.

I

Impermeability: the incapacity of a rock to transmit a fluid.

Index of biotic integrity (IBI): an aggregated number, or index, based on several attributes or metrics of a fish community that provides an assessment of biological conditions.

Indicator sites: stream sampling sites located at outlets of drainage basins with relatively homogeneous land use and physiographic conditions; most indicator-site basins have drainage areas ranging from 20 to 200 square miles.

Infiltration: the downward movement of water from the atmosphere into the soil or porous rock.

Influent: Water flowing into a reservoir, basin, or treatment plant.

Inorganic: containing no carbon; matter other than plant or animal.

Inorganic chemical: a chemical substance of mineral origin not having carbon in its molecular structure.

Inorganic soil: soil with less than 20% organic matter in the upper 16 inches.

Ionic bond: the attractive forces between oppositely charged ions—for example, the forces between the sodium and chloride ions in a sodium chloride crystal.

Instantaneous discharge: the volume of water that passes a point at a particular instant of time.

Instream use: water use taking place within the stream channel for such purposes as hydroelectric power generation, navigation, water-quality improvement, fish propagation, and recreation. Sometimes called nonwithdrawal use or in-channel use.

Intermittent stream: a stream that flows only when it receives water from rainfall-runoff or springs, or from some surface source such as melting snow.

Internal drainage: surface drainage whereby the water does not reach the ocean, such as drainage toward the lowermost or central part of an interior basin or closed depression.

Intertidal: alternately flooded and exposed by tides.

Intolerant organisms: organisms that are not adaptable to human alterations to the environment and thus decline in numbers where alterations occur.

Invertebrate: an animal having no backbone or spinal column.

Ion: a positively or negatively charged atom or group of atoms.

Irrigation: controlled application of water to arable land to supply requirements of crops not satisfied by rainfall.

Irrigation return flow: the part of irrigation applied to the surface that is not consumed by evapotranspiration or uptake by plants and that migrates to an aquifer or surface-water body.

Irrigation withdrawals: withdrawals of water for application on land to assist in the growing of crops and pastures or to maintain recreational lands.

K

Karst: a type of topography that is formed on limestone, dolomite, gypsum, and other rocks, primarily by dissolution, and that is characterized by sinkholes, caves, and subterranean drainage.

Karst topography: type of topography that is formed in limestone, gypsum, and other similar type rock by dissolution and is characterized by sinkholes, caves, and rapid underground water movement.

Karstification: action by water, mainly chemical but also mechanical, that produces features of a karst topography.

Karst mantled: a terrane of karst features, usually subdued, and covered by soil or a thin alluvium.

Kill: Dutch term for stream or creek.

L

Lacustrine: pertaining to, produced by, or formed in a lake.

Leachate: a liquid that has percolated through soil containing soluble substances and that contains certain amounts of these substances in solution.

Leaching: the removal of materials in solution from soil or rock; also refers to the movement of pesticides or nutrients from land surface to groundwater.

Limnetic: the deep-water zone (greater than 2 meters deep).

Littoral: the shallow-water zone (less than 2 meters deep).

Load: material that is moved or carried by streams, reported as the weight of material transported during a specified time period, such as tons per year.

M

Main stem: the principal trunk of a river or a stream.

Marsh: a water-saturated, poorly drained area, intermittently or permanently water covered, having aquatic and grass-like vegetation.

Maturity (stream): the stage in the development of a stream at which it has reached its maximum efficiency, when velocity is just sufficient to carry the sediment delivered to it by tributaries; characterized by a broad, open, flat-floored valley having a moderate gradient and gentle slope.

Maximum contaminant level (MCL): a primary standard, whereas an MCLG is a maximum concentration goal for a drinking water contaminant, which would be desirable on human health concerns and assuming all feasibility issues such as cost and technological capacity are not considered. Stated differently, MCL is the maximum allowable concentration of a contaminant in drinking water, as established by state and/or federal regulations. Primary MCLs are health related and mandatory. Secondary MCLs are related to the aesthetics of the water and are highly recommended, but not required.

Mean discharge: the arithmetic mean of individual daily mean discharges of a stream during a specific period, usually daily, monthly, or annually.

Membrane filter method: a laboratory method used for coliform testing. The procedure uses an ultra-thin filter with a uniform pore size smaller than bacteria (less than a micron). After the water is forced through the filter, the filter is incubated in a special media that promotes the growth of coliform bacteria. Bacterial colonies with a green-gold sheen indicate the presence of coliform bacteria.

Method detection limit: the minimum concentration of a substance that can be accurately identified and measured with current lab technologies.

Midge: a small fly in the family Chironomidae. The larval (juvenile) life stages are aquatic.

Minimum reporting level (MRL): the smallest measured concentration of a constituent that may be reliably reported using a given analytical method. In many cases, the MRL is used when documentation for the method detection limit is not available.

Mitigation: actions taken to avoid, reduce, or compensate for the effects of human-induced environmental damage.

Modes of transmission of disease: the ways in which diseases are spread from one person to another.

Monitoring: repeated observation, measurement, or sampling at a site, on a scheduled or event basis, for a particular purpose.

Monitoring well: a well designed for measuring water levels and testing groundwater quality.

Multiple-tube fermentation method: a laboratory method used for coliform testing, which uses a nutrient broth placed in a culture tube. Gas production indicates the presence of coliform bacteria.

N

National Primary Drinking Water Regulations (NPDWRs): regulations developed under the Safe Drinking Water Act, which establish maximum contaminant levels, monitoring requirements, and reporting procedures for contaminants in drinking water that endanger human health.

National pollutant discharge elimination system (NPDES): a requirement of the CWA that discharges meet certain requirements prior to discharging waste to any water body. It sets the highest permissible effluent limits, by permit, prior to making any discharge.

Near coastal water initiative: this initiative was developed in 1985 to provide for the management of specific problems in waters near coastlines that are not dealt with in other programs.

Nitrate: an ion consisting of nitrogen and oxygen (NO_3). Nitrate is a plant nutrient and is very mobile in soils.

Nonbiodegradable: substances that do not break down easily in the environment.

Nonpoint source: a source (of any water-carried material) from a broad area, rather than from discrete points.

Nonpoint-source contaminant: a substance that pollutes or degrades water that comes from lawn or cropland runoff, the atmosphere, roadways, and other diffuse sources.

Nonpoint-source water pollution: water contamination that originates from a broad area (such as leaching of agricultural chemicals from a cropland) and enters the water resource diffusely over a large area.

Nonpolar covalently bonded: a molecule composed of atoms that share their electrons equally, resulting in a molecule that does not have polarity.

Nutrient: any inorganic or organic compound needed to sustain plant life.

O

Organic: containing carbon, but possibly also containing hydrogen, oxygen, chlorine, nitrogen, and other elements.

Organic chemical: a chemical substance of animal or vegetable origin having carbon in its molecular structure.

Organic detritus: any loose organic material in streams—such as leaves, bark, or twigs—removed and transported by mechanical means, such a disintegration or abrasion.

Organic soil: soil that contains more than 20% organic matter in the upper 16 inches.

Organochlorine compound: synthetic organic compounds containing chlorine. As generally used, this term refers to compounds containing mostly or exclusively carbon, hydrogen, and chlorine.

Outwash: soil material washed down a hillside by rainwater and deposited upon more gently sloping land.

Overdraft: any withdrawal of groundwater in excess of the *Safe Yield.*

Overland flow: the flow of rainwater or snowbelt over the land surface toward stream channels.

Oxidation: when a substance either gains oxygen or loses hydrogen or electrons in a chemical reaction. One of the chemical treatment methods.

Oxidizer: a substance that oxidizes another substance.

P

Paleokarst: a karstified area that has been buried by later deposition of sediments.

Parts per million: the number of weight or volume units of a constituent present with each one million units of the solution or mixture. Formerly used to express the results of most water and wastewater analyses, **PPM** is being replaced by milligrams per liter **M/L**. For drinking water analyses, concentration in parts per million and milligrams per liter are equivalent. A single PPM can be compared to a shot glass full of water inside a swimming pool.

Pathogens: types of microorganisms that can cause disease.

Perched groundwater: unconfined groundwater separated from an underlying main body of groundwater by an unsaturated zone.

Percolation: the movement, under hydrostatic pressure, of water through interstices of a rock or soil (except the movement through large openings such as caves).

Perennial stream: a stream that normally has water in its channel at all times.

Periphyton: microorganisms that coat rocks, plants, and other surfaces on lake bottoms.

Permeability: the capacity of a rock for transmitting a fluid; a measure of the relative ease with which a porous medium can transmit a liquid. The quality of the soil that enables water to move downward through the soil profile. Permeability is measured as the number of inches per hour that water moves downward through the saturated soil. Terms describing permeability are:

Very slow	less than 0.06 inches/hr
Slow	0.06–0.2 inches/hr
Moderately slow	0.2–0.6 inches/hr
Moderate	0.6–2.0 inches/hr
Moderately rapid	2.0–6.0 inches/hr
Rapid	6.0–20 inches//hr
Very Rapid	more than 20 inches/hr

pH: a measure of the acidity (less than 7) or alkalinity (greater than 7) of a solution; a pH of 7 is considered neutral.

Phosphorus: a nutrient essential for growth that can play a key role in stimulating aquatic growth in lakes and streams.

Photosynthesis: the synthesis of compounds with the aid of light.

Physical treatment: any process that does not produce a new substance (e.g., screening, adsorption, aeration, sedimentation, and filtration).

Plutonic: a loosely defined term with a number of current usages. We use it to describe igneous rock bodies that crystallized at great depth or, more generally, any intrusive igneous rock.

Point source: originating at any discrete source.

Polar covalent bond: the shared pair of electrons between two atoms are not equally held. Thus, one of the atoms becomes slightly positively charged and the other atom becomes slightly negatively charged.

Polar covalent molecule: (water) one or more polar covalent bonds result in a molecule that is polar covalent. Polar covalent molecules exhibit partial positive and negative poles, causing them to behave like tiny magnets. Water is the most common polar covalent substance.

Pollutant: any substance introduced into the environment that adversely affects the usefulness of the resource.

Pollution: the presence of matter or energy whose nature, location, or quantity produces undesired environmental effects. Under the Clean Water Act, for example, the term is defined as a man-made or man-induced alteration of the physical, biological, and radiological integrity of water.

Polychlorinated biphenyls (PCBs): a mixture of chlorinated derivatives of biphenyl, marketed under the trade name Aroclor with a number designating the chlorine content (such as Aroclor 1260). PCBs were used in transformers and capacitors for insulating purposes and in gas pipeline systems as a lubricant. The further sale or new use was banned by law in 1979.

Polycyclic aromatic hydrocarbon (PAH): a class of organic compounds with a fused-ring aromatic structure. PAHs result from incomplete combustion of organic carbon (including wood), municipal solid waste, and fossil fuels, as well as from natural or anthropogenic introduction of uncombusted coal and oil. PAHs included benzo(a)pyrene, fluoranthene, and pyrene.

Population: a collection of individuals of one species or mixed species making up the residents of a prescribed area.

Porosity: (1) The ratio of the aggregate volume of pore spaces in rock or soil to its total volume, usually stated as a percent. (2) A measure of the water-bearing capacity of subsurface rock. With respect to water movement, it is not just the total magnitude of porosity that is important, but the size of the voids and the extent to which they are interconnected, as the pores in a formation may be open, or interconnected, or closed and isolated. For example, clay may have a very high porosity with respect to potential water content, but it constitutes a poor medium as an aquifer because the pores are usually so small.

Potable water: water that is safe and palatable for human consumption.

Potentiometric surface: a surface that represents the total head in an aquifer; that is, it represents the height above a datum plane at which the water level stands in tightly cased wells that penetrated the aquifer.

Precipitation: any or all forms of water particles that fall from the atmosphere, such as rain, snow, hail, and sleet. The act or process of producing a solid phase within a liquid medium.

Pretreatment: any physical, chemical, or mechanical process used before the main water treatment processes. It can include screening, presedimentation, and chemical addition.

Primary drinking water standards: regulations on drinking water quality (under SWDA) considered essential for the preservation of public health.

Primary treatment: the first step of treatment at a municipal wastewater treatment plant. It typically involves screening and sedimentation to remove materials that float or settle.

Public-supply withdrawals: water withdrawn by public and private water suppliers for use within a general community. Water is used for a variety of purposes such as domestic, commercial, industrial, and public water use.

Public water system: as defined by the Safe Drinking Water Act, any system, publicly or privately owned, that serves at least 15 service connections 60 days out of the year or serves an average of 25 people at least 60 days out of the year.

Publicly owned treatment works (POTW): a waste treatment works owned by a state, local government unit, or Indian tribe, usually designed to treat domestic wastewaters.

R

Rain shadow: a dry region on the lee side of a topographic obstacle, usually a mountain range, where rainfall is noticeably less than on the windward side.

Reach: a continuous part of a stream between two specified points.

Reaeration: the replenishment of oxygen in the water from which oxygen has been removed.

Receiving waters: a river, lake, ocean, stream, or other water sources into which wastewater or treated effluent is discharged.

Recharge: the process by which water is added to a zone of saturation, by percolation from the soil surface or by artificial injection.

Recharge area (groundwater): an area within which water infiltrates the ground and reaches the zone of saturation.

Reference dose (RfD): an estimate of the amount of a chemical that a person can be exposed to on a daily basis that is not anticipated to cause adverse systemic health effects over the person's lifetime.

Representative sample: a sample containing all the constituents present in the water from which it was taken.

Return flow: that part of irrigation water that is not consumed by evapotranspiration and that returns to its source or another body of water.

Reverse osmosis (RO): solutions of differing ion concentration are separated by a semipermeable membrane. Typically, water flows from the chamber with lesser ion concentration into the chamber with the greater ion concentration, resulting in hydrostatic or osmotic pressure. In RO, enough external pressure is applied to overcome this hydrostatic pressure, thus reversing the flow of water. This results in the water on the other side of the membrane becoming depleted in ions and demineralized.

Riffle: a shallow part of the stream where water flows swiftly over completely or partially submerged obstructions to produce surface agitation.

Riparian: pertaining to or situated on the bank of a natural body of flowing water.

Riparian rights: a concept of water law under which authorization to use water in a stream is based on ownership of the land adjacent to the stream.

Riparian zone: pertaining to or located on the bank of a body of water, especially a stream.

Rock: any naturally formed, consolidated, or unconsolidated material (but not soil) consisting of two or more minerals.

Runoff: that part of precipitation or snowmelt that appears in streams or surface-water bodies.

Rural withdrawals: water used in suburban or farm areas for domestic and live-stock needs. The water generally is self-supplied and includes domestic use, drinking water for livestock, and other uses such as dairy sanitation, evaporation from stock-watering ponds, and cleaning and waste disposal.

S

Safe Drinking Water Act (SDWA): a federal law passed in 1974 with the goal of establishing federal standards for drinking water quality, protecting underground sources of water, and setting up a system of state and federal cooperation to assure compliance with the law.

Saline water: water that is considered unsuitable for human consumption or for irrigation because of its high content of dissolved solids; generally expressed as milligrams per liter (mg/L) of dissolved solids; seawater is generally considered to contain more than 35,000 mg/L of dissolved solids. A general salinity scale is:

	Concentration of dissolved solids in mg/L
Slightly saline	1,000–3,000
Moderately saline	3,000–10,000
Very saline	10,000–35,000
Brine	More than 35,000

Saturated zone: a subsurface zone in which all the interstices or voids are filled with water under pressure greater than that of the atmosphere.

Screening: a pretreatment method that uses coarse screens to remove large debris from the water to prevent clogging of pipes or channels to the treatment plant.

Secondary drinking water standards: regulations developed under the Safe Drinking Water Act that established maximum levels of substances affecting the aesthetic characteristics (taste, color, or odor) of drinking water.

Secondary maximum contaminant level (SMCL): the maximum level of a contaminant or undesirable constituent in public water systems that, in the judgment of US EPA, is required to protect the public welfare. SMCLs are secondary (nonenforceable) drinking water regulations established by the US EPA for contaminants that may adversely affect the odor or appearance of such water.

Secondary treatment: the second step of treatment at a municipal wastewater treatment plant. This step uses growing numbers of microorganisms to digest organic matter and reduce the amount of organic waste. Water leaving this process is chlorinated to destroy any disease-causing microorganisms before its release.

Sedimentation: a physical treatment method that involves reducing the velocity of water in basins so that the suspended material can settle out by gravity.

Seep: a small area where water percolates slowly to the land surface.

Seiche: a sudden oscillation of the water in a moderate-size body of water, caused by wind.

Sinkhole: a depression in a karst area. At the land surface, its shape is generally circular and its size measured in meters to tens of meters; underground it is commonly funnel-shaped and associated with subterranean drainage.

Sinuosity: the ratio of the channel length between two points on a channel to the straight-line distance between the same two points; a measure of meandering.

Soil: the layer of material at the land surface that supports plant growth.

Soil horizon: a layer of soil that is distinguishable from adjacent layers by characteristic physical and chemical properties.

Soil moisture: water occurring in the pore spaces between the soil particles in the unsaturated zone from which water is discharged by the transpiration of plants or by evaporation from the soil.

Solution: formed when a solid, gas, or another liquid in contact with a liquid becomes dispersed homogeneously throughout the liquid. The substance called a solute is said to dissolve. The liquid is called the solvent.

Solvated: when either a positive or negative ion becomes completely surrounded by polar solvent molecules.

Sorb: to take up and hold either by absorption or adsorption.

Sorption: general term for the interaction (binding or association) of a solute ion or molecule with a solid.

Spall: a chip or fragment removed from a rock surface by withering, especially by the process of exfoliation.

Specific capacity: the yield of a well per unit of drawdown.

Specific retention: the ratio of the volume of water retained in a rock after gravity drainage to the volume of the rock.

Specific storage: the volume of water that an aquifer system releases or takes into storage per unit volume per unit change in head. The specific storage is equivalent to the *Storage Coefficient* divided by the thickness of the aquifer system.

Specific yield: the ratio of the volume of water that will drain under the influence of gravity to the volume of saturated rock.

Spring: place where any natural discharge of groundwater flows at the ground surface.

Storage: the capacity of an aquifer, aquitard, or aquifer system to release or accept water into groundwater storage, per unit change in hydraulic head.

Storage coefficient: the volume of water released from storage in a unit prism of an aquifer when the head is lowered a unit distance.

Strain: Deformation that results from stress. Expressed in terms of the amount of deformation per inch.

Stratification: the layered structure of sedimentary rocks.

Stress and strain: in materials, stress is a measure of the deforming force applied to a body. Strain (which is often erroneously used as a synonym for stress) is really the resulting change in its shape (deformation). For a perfectly elastic material, stress is proportional to strain. This relationship is explained by Hooke's Law, which states that the deformation of a body is proportional to the magnitude of the deforming force, provided that the body's elastic limit is not exceeded. If the elastic limit is not reached, the body will return to its original size once the force is removed. For example, if a spring is stretched by 2 cm by a weight of 1 N, it will be stretched by 4 cm by a weight of 2 N, and so on; however, once the load exceeds the elastic limit for the spring, Hooke's law will no longer be obeyed, and each successive increase in weight will result in a greater extension until the spring finally breaks.

Stress forces are categorized in three ways:

1. Tension (or tensile stress), in which equal and opposite forces that act away from each other are applied to a body; tends to elongate the body.
2. Compression stress, in which equal and opposite forces that act toward each other are applied to a body; tends to shorten it.
3. Shear stress, in which equal and opposite forces that do not act along the same line of action or plane are applied to a body; tends to change its shape without changing its volume.

Stress, geostatic (lithostatic): the total weight (per unit area) of sediments and water above some plane of reference. Geostatic stress normal to any horizontal plane of reference in a saturated deposit may also be defined as the sum of the effective stress and the fluid pressure at that depth.

Stress, preconsolidation: the maximum antecedent effective stress to which a deposit has been subjected and which it can withstand without undergoing additional permanent deformation. Stress changes in the range less than the preconsolidation stress produce elastic deformations of small magnitude. In fine-grained materials, stress increase beyond the preconsolidation stress produced much larger deformations that are principally inelastic (nonrecoverable). Synonymous with "virgin stress."

Stress, seepage: force (per unit area) transferred from the water to the medium by viscous friction when water flows through a porous medium. The force transferred to the medium is equal to the loss of hydraulic head and is termed the seepage force exerted in the friction of flow.

Subsidence: a dropping of the land surface as a result of groundwater being pumped. Cracks and fissures can appear in the land. Some state that subsidence is virtually an irreversible process. Others, like the author of this book, states that the jury is still out on the validity of this statement; HRSD's SWIFT project may prove that land subsidence can be reversed.

Surface runoff: runoff that travels over the land surface to the nearest stream channel.

Surface tension: the attractive forces exerted by the molecules below the surface upon those at the surface, resulting in them crowding together and forming a higher density.

Surface water: all water naturally opens to the atmosphere, and all springs, wells, or other collectors that are directly influenced by surface water.

Surface Water Treatment Rule (SWTR): a federal regulation established by the USEPA under the Safe Drinking Water Act that imposes specific monitoring and treatment requirements on all public drinking water systems that draw water from a surface water source.

Suspended sediment: sediment that is transported in suspension by a stream.

Suspended solids: different from suspended sediment only in the way that the sample is collected and analyzed.

Synthetic organic chemicals (SOCs): generally applied to manufactured chemicals that are not as volatile as volatile organic chemicals. Included are herbicides, pesticides, and chemicals widely used in industries.

T

Total head: the height above a datum plane of a column of water. In a groundwater system, it is composed of an elevation head and pressure head.

Total suspended solids (TSS): solids present in wastewater.

Transmissivity (groundwater): The capacity of a rock to transmit water under pressure. The coefficient of transmissibility is the rate of flow of water, at the prevailing water temperature, in gallons per day, through a vertical strip of the aquifer one foot wide, extending the full saturated height of the aquifer under a hydraulic gradient of 100-percent. A hydraulic gradient of 100-percent means a one-foot drop in head in one foot of flow distance.

Transpiration: the process by which water passes through living organisms, primarily plants, into the atmosphere.

Trihalomethanes (THMs): a group of compounds formed when natural organic compounds from decaying vegetation and soil (such as humic and fulvic acids) react with chlorine.

Turbidity: a measure of the cloudiness of water caused by the presence of suspended matter, which shelters harmful microorganisms and reduces the effectiveness of disinfecting compounds.

U

Unconfined aquifer: an aquifer whose upper surface is a water table free to fluctuate under atmospheric pressure.

Unsaturated zone: a subsurface zone above the water table in which the pore spaces may contain a combination of air and water.

V

Vehicle of disease transmission: any nonliving object or substance contaminated with pathogens.

Vernal pool: a small lake or pond that is filled with water for only a short time during spring.

Vug: a small cavity or chamber in rock that may be lined with crystals.

W

Wastewater: the spent or used water from individual homes, a community, a farm, or an industry that contains dissolved or suspended matter.

Water budget: an accounting of the inflow to, outflow from, and storage changes of water in a hydrologic unit.

Water column: an imaginary column extending through a water body from its floor to its surface.

Water demand: water requirements for a particular purpose, such as irrigation, power, municipal supply, plant transpiration, or storage.

Water table: the topwater surface of an unconfined aquifer at atmospheric pressure.

Waterborne disease: water is a potential vehicle of disease transmission, and waterborne disease is possibly one of the most preventable types of communicable illness. The application of basic sanitary principles and technology have virtually eliminated serious outbreaks of waterborne diseases in developed countries. The most prevalent waterborne diseases include *typhoid fever, dysentery, cholera, infectious hepatitis,* and *gastroenteritis.*

Water softening: a chemical treatment method that uses either chemicals to precipitate or a zeolite to remove those metal ions (typically Ca^{2+}, Mg^{2+}, Fe^{3+}) responsible for hard water.

Watershed: the land area that drains into a river, river system, or other body of water.

Wellhead protection: the protection of the surface and subsurface areas surrounding a water well or well field supplying a public water system from contamination by human activity.

Y

Yield: the mass of material or constituent transported by a river in a specified period of time divided by the drainage area of the river basin.

Yield, optimal: an optimal amount of groundwater, by virtue of its use, that should be withdrawn from an aquifer system or groundwater basin each year. It is a dynamic quantity that must be determined from a set of alternative groundwater management decisions subject to goals, objectives, and constraints of the management plan.

Yield, perennial: the amount of usable water from an aquifer that can be economically consumed each year for an indefinite period of time. It is a specified amount that is commonly specified equal to the mean annual recharge to the aquifer system, which thereby limits the amount of groundwater that can be pumped for beneficial use.

Yield, safe: the amount of groundwater that can be safely withdrawn from a groundwater basin annually, without producing an undesirable result. Undesirable results include but are not limited to depletion of groundwater storage, the intrusion of water of undesirable quality, the contraventions of existing water rights, the deterioration of the economic advantages of pumping (such as excessively lower water levels and the attendant increased pumping lifts and associated energy costs), excessive depletion of stream flow by induced infiltration, and land subsidence.

Z

Zone of aeration: the zone above the water table. Water in the zone of aeration does not flow into a well.

Zone of capillarity: the area above a water table where some or all of the interstices (pores) are filled with water that is held by capillarity.

Index

Printed in the United States
by Baker & Taylor Publisher Services